Pharmacogenetics: Opportunities and Challenges for Health Innovation

OECD

ORGANISATION FOR ECONOMIC CO-OPERATION AND DEVELOPMENT

The OECD is a unique forum where the governments of 30 democracies work together to address the economic, social and environmental challenges of globalisation. The OECD is also at the forefront of efforts to understand and to help governments respond to new developments and concerns, such as corporate governance, the information economy and the challenges of an ageing population. The Organisation provides a setting where governments can compare policy experiences, seek answers to common problems, identify good practice and work to co-ordinate domestic and international policies.

The OECD member countries are: Australia, Austria, Belgium, Canada, the Czech Republic, Denmark, Finland, France, Germany, Greece, Hungary, Iceland, Ireland, Italy, Japan, Korea, Luxembourg, Mexico, the Netherlands, New Zealand, Norway, Poland, Portugal, the Slovak Republic, Spain, Sweden, Switzerland, Turkey, the United Kingdom and the United States. The Commission of the European Communities takes part in the work of the OECD.

OECD Publishing disseminates widely the results of the Organisation's statistics gathering and research on economic, social and environmental issues, as well as the conventions, guidelines and standards agreed by its members.

ISBN 978-92-64-07679-2 (print)
ISBN 978-92-64-07680-8 (PDF)

Corrigenda to OECD publications may be found on line at: *www.oecd.org/publishing/corrigenda*.

Foreword

Pharmacogenetics offers new ways of understanding how medicines work as well as their safety and efficacy in individuals. This can lead to a more efficient and effective approach to drug discovery. In addition, pharmacogenetics may lead to a more diversified and targeted portfolio of diagnostics and therapies, which, when used together, would yield greater health benefits. This book examines the present use and future challenges facing pharmacogenetics at different stages of the health innovation cycle, from research through to its uptake in the clinic. The report reviews the evidence to date as to the impact of pharmacogenetics on decision making and efficiency in pharmaceutical R&D and in clinical care.

This publication also identifies concrete policies governments need to put in place in order to facilitate the uptake of this approach to R&D and clinical care, and to maximise its social benefits. A number of scientific, economic and regulatory challenges need to be overcome if pharmacogenetics is to be taken up more widely within health-care systems. It is not yet clear to the private sector what business models will deliver acceptable returns on investment, but in the public sector, governments have a number of levers which they could use to create an "enabling" environment for pharmacogenetics, while continuing to provide the necessary checks and balances.

This report draws on debates first held under the auspices of the OECD Working Party on Biotechnology (WPB) at a workshop entitled "An International Perspective on Pharmacogenetics: The Intersections between Innovation, Regulation and Health Delivery" in Rome, Italy, in 2005. As it developed, the report benefitted from comments from experts at the workshop and from delegates to the Working Party on Biotechnology. Several outside experts have provided important written contributions, including Celia Caulcott, Peter Greenaway and Kees van Gool.

The report was edited by Marilyn Smith and Lantz Miller. In the OECD Secretariat, Elettra Ronchi was responsible for the project and compiled consultant reports, while Bénédicte Callan supervised the final drafting and editing. Stella Horsin provided critical administrative support.

Table of Contents

Acronyms

ADMET (ADME-Tox)	Absorption, diffusion, metabolism, elimination, toxicology
ADRs	Adverse drug reactions
BLA	Biologic license applications
CEA	Cost-effectiveness analysis
CYP450	Cytochrome P450 gene, involved in metabolism of certain medecines
DNA	Deoxyribonucleic acid
Dx	Diagnostic test
EFPIA	European Federation of Pharmaceutical Industries and Associations
EMEA	European Medecines Agency
FDA	US Federal Drug Administration
HBGRDs	Human biobanks and genetic research databases
HER2	Human epidermal growth factor receptor-2, gene associated with some types of breast cancer
ICER	Incremental cost-effectiveness ratio
IMI	Innovative Medicines Initiative
IND	Investigational new drug
NIH	National Institutes of Health (United States)
NME	New Molecular Entity
OECD	Organisation for Economic Co-operation and Development
R&D	Research and development
PCR	Polymerase chain reaction
PGx	Pharmacogenetics
PhRMA	Pharmaceutical Research and Manufacturers of America

QALYs	Quality-adjusted life years
siRNA	Small interfering ribonucleic acid
SMEs	Small and medium sized enterprises
SNPs	Single nucleotide polymorphisms
TPMT	Thiopurine methyltransferase, gene involved in the metabolism of two compounds used in a number of medicines
Tx	Therapeutic agent
USD	US dollars
VGDS	Voluntary genomic data submission

Executive Summary

Pharmacogenetics offers new ways of understanding how drugs work and how this affects both the safety and efficacy of drugs in individuals. The potential opportunities are considerable. In drug development, pharmacogenetics shows great potential for improving the efficiency of the drug discovery process, particularly for identifying and validating new drug targets. It is expected to improve the translation of early-stage projects into medicines that meet public health needs. In clinical care, pharmacogenetics may enable doctors to prescribe more effective interventions and improve their use of evidence-based medicine. It can help identify those individuals most likely to benefit from a therapy, thereby optimising treatment strategies for both common and complex disorders. It may also enable preventive interventions.

This book examines the present use and future challenges facing pharmacogenetics at different stages in the health innovation cycle, including its uptake in the clinic. The report draws on debates held under the auspices of the OECD's Working Party on Biotechnology (WPB) which were initiated at an OECD workshop entitled "An International Perspective on Pharmacogenetics: the Intersections between Innovation, Regulation and Health Delivery", held in Rome, Italy, in 2005. The report reviews the evidence to date as to the impact of pharmacogenetics on decision making and efficiency in pharmaceutical R&D and in clinical care. Finally, it identifies policies governments need to put in place in order to facilitate the uptake of this approach to R&D and clinical care and to maximise its benefits.

The value of pharmacogenetics is heavily dependent on the identification of useful "biomarkers". Biomarkers are indicators that mark the presence of a potential gene-drug interaction or that measure response to therapeutic activity. Genetic biomarkers – which identify genetic variations in patient populations – are emerging as one of the most effective means of improving the efficiency of the drug discovery process. They can be used in clinical trials to stratify patients who respond to a new potential medicine appropriately, adversely or not at all. In clinical practice, genetic biomarkers can be developed into diagnostic tests to measure the potential efficacy or toxicity of a particular therapy. For medicines that eventually are brought to market for a genotype-specific population, these tests may be necessary to define which patients will benefit from the treatment and which will not. Pharmacogenetics is already being employed in a small number of cases to optimise treatment strategies for common and complex disorders of public health relevance.

The use of pharmacogenetics is progressing rapidly. Its impacts are evident in three areas: *i)* basic research; *ii)* drug discovery and development; and *iii)* management of health care. But while research in pharmacogenetics is proceeding apace, by early 2009 only a few pharmacogenetics-based diagnostics were on the market. In fact, six years after the initial sequencing of the human genome, fewer than a dozen new pharmacogenetic products are commercially available. A number of scientific, regulatory and economic challenges need to be overcome if pharmacogenetics is to be taken up more widely by health-care systems.

The report concludes that governments have a role to play in creating an "enabling" environment for the uptake of pharmacogenetics. Six key messages aim to guide elaboration of public policy and co-ordinated international action.

---

1. Building infrastructures for large-scale association studies is necessary to identify and validate the biomarkers that underpin the use of pharmacogenetics.

Much work must be done to identify a biomarker, carry out studies to verify its association with relevant health outcomes or therapy-related effects, and validate its use as a diagnostic tool for clinical practice. Identifying and ultimately validating biomarkers requires integrating genetic and genomic data with phenotypic data. This can be both difficult and expensive because it requires accessing and integrating different types of data at multi-scale levels (from the molecular to the clinical) and in different formats.

Moreover, association studies usually involve clinical studies with a large number of patients, often from a variety of population groups. To carry out large-scale association studies, appropriate frameworks, systems and methodologies must be established.

Governments may be able to facilitate this process by helping to build the required research infrastructure. They might:

- encourage the formation of multidisciplinary international networks that can facilitate access to the necessary data sources and increase the efficiency of pharmacogenetic research.
- encourage agreements relating to the availability of raw data and the sharing of data.

- support the creation and utilisation of large-scale human biobanks and genetic research databases (HBGRDs).
- consider the formation of public-private partnerships to carry out association studies.
- foster the development of systems to manage knowledge and intellectual property so as to support open innovation platforms for pharmacogenetics.

2. Applying pharmacogenetics to established medicines will yield public health benefits but will require public as well as private support and collaboration in order to run the necessary prospective studies.

Frequently, adverse drug reactions are observed only once medicines have been on the market for some time and many thousands of patients have been exposed. The incentives for pharmaceutical or devices companies to apply pharmacogenetics to established medicines (particularly those that are off-patent) are weak, even though substantial benefits would accrue to patients and society as a whole in the form of reduced adverse drug reactions.

Applying pharmacogenetics to existing and common drugs requires major prospective studies to identify relevant genetic markers for patient stratification. Such prospective studies would allow regulatory authorities to determine whether it is necessary to modify the labelling of specific established medicines in order to improve clinical practice and patient health. However, as the cost of such trials may be very high, the application of pharmacogenetics to established medicines becomes an issue of public policy.

3. Pharmacogenetics has the potential to transform the drug development process, but incentives to adopt this technology may need to be strengthened.

Pharmacogenetics can be used to reduce the size, duration and cost of clinical trials. However, the adoption of pharmacogenetics will pose economic and organisational challenges for industry as it may entail substantial reforms both of the drug discovery process and of the business models for pharmaceutical and diagnostic firms. For the pharmaceutical industry, to the extent that pharmacogenetics reduces the size of the population for any given drug, it puts pressure on the blockbuster business model. For the devices and diagnostic industries, uncertainties include how to capture value from assays, and from challenges relating to co-developing and co-marketing assays and therapeutics.

Drug and device development has occurred independently until now. In pharmacogenetics, drugs and diagnostics could be co-developed and co-marketed: the need to co-ordinate and synchronise development may encourage pharmaceutical companies to partner with device companies or seek to develop their own in-house expertise.

The economic incentives to invest in the development of biomarkers are influenced by signals that flow from the broader health-care system. Within current pricing and reimbursement mechanisms, the lack of recognition of the value of testing represents a disincentive for the devices industry to develop new, genetics-based assays. Policies can help make the uptake of pharmacogenetics more attractive. Clear signals from governmental and regulatory bodies about how this technology will be priced and reimbursed, recognition of the added value of diagnostic tests for the health system as a whole, and mechanisms for capturing and protecting the intellectual property inherent in diagnostic tests might improve the incentives for investment.

4. Co-ordination and dialogue with regulatory authorities are critical to strengthening investments in the development of pharmacogenetic products by industry.

Regulation of the combined use (co-development) of a therapeutic with a diagnostic is evolving across OECD countries. There is concern that pharmacogenetics will make an already complex approval process for pharmaceuticals even more complicated. Clarity about how regulation will deal with the co-development issue is therefore necessary, in terms both of the data requirements for approval and of how reimbursement systems will react to, and value, the co-marketing of co-developed products.

When evaluating new pharmacogenetic drugs and associated tests, decision makers will want to balance benefits and needs and find evidence of value for money. The use of pharmacogenetic testing for the prescription of new drugs may make them more expensive. However, the costs of inappropriate prescribing or the results of the adverse reactions that will be prevented/reduced may more than offset the added costs of the pharmacogenetic tests, resulting in increased clinical value to the patient and prescriber.

To reduce uncertainties, innovators, regulators and end users may need to engage in dialogue to clarify how the approval process and subsequent reimbursement/coverage decisions might proceed for pharmacogenetic products. OECD countries may, for example, want to consider the options of conditional market approvals, through the use of post-market pharmacovigilance, and risk-sharing mechanisms for pharmacogenetic products. The development of common methodologies or approaches to policies, coupled with efforts to close existing gaps in regulatory and technology assessment practices could also help improve incentives to innovate across the OECD.

5. The health and economic impacts of pharmacogenetics need to be better understood if the health-care system is to adopt this new technology.

As with any health-care innovation, introducing pharmacogenetic testing into health care will pose a number of challenges for health-care systems. Studies on the health economics of pharmacogenetics and on the cost-benefit ratio of pharmacogenetic testing and products could provide the evidence base necessary for their uptake. Presently there is a lack of data demonstrating the clinical utility and cost-effectiveness of many pharmacogenetic therapeutics and diagnostics, and no agreement over whose responsibility it is to develop such data.

Although existing health technology assessment models for evaluating new genetic tests and medicines are generally regarded as acceptable, this may not be the case in future if the numbers of pharmacogenetic products increases significantly. There is likely to be a need to develop new models and methodologies for the assessment of diagnostics and medicines which may eventually influence the pricing and reimbursement of such products.

6. Health-care providers will need to be educated about pharmacogenetic assays and treatment options, and they must have easy access to clinically useful information at the point of care in order to interpret these assays.

In health-care decision making, pharmacogenetics is contributing to better clinical care. It is improving our understanding of disease heterogeneity, reducing the uncertainties of responses associated with specific treatments, changing the risk-benefit ratio for treatments, and enhancing the ability to prescribe accurate dosage for medicines.

However, pharmacogenetic assays rarely provide simple, clear-cut results: the information generated is probabilistic rather than absolute. Before pharmacogenetic testing becomes a routine part of care, there needs to be more evidence to support the clinical utility of pharmacogenetic testing, information on pharmacogenetics and other relevant information need to be available at the point of care, and health-care providers need to be educated and trained to access and interpret these new sources of health data.

Chapter 1

Introduction to the Policy issues

This chapter serves as a general introduction to the issues raised by the potential adoption of pharmacogenetics in drug development and delivery, including its economic implications, which are examined in greater detail in the rest of the report.

Introduction

Advances in human genomic research, coupled with breakthroughs in fields that span biomedical as well as computational biology, are resulting in greater understanding of the natural history of disease and of the causative mechanisms involved. Collectively, these advances may radically transform the practice of medicine, the prevention, diagnosis and treatment of disease, and the delivery of health-care services.

So dynamic are these developments that in 2004 OECD member country Ministers of Science and Technology[1] and Ministers of Health recognised the need to strengthen our understanding of human genomics and promote its use. In particular they thought that genomics should be employed to help achieve the dual goals of generating economic growth and delivering better health outcomes. The ministers concluded that policies promoting genomic innovations should aim at broad-based public health benefits as well as improvements in the health innovation process and voiced concern over whether this would be the case without further adjustments.

The field of pharmacogenetics is one of the first commercially important outgrowths of the Human Genome Project and major related genomic research undertakings. Pharmacogenetics focuses on how individuals' genetic profiles correspond to the variable ways in which humans respond to pharmaceuticals. Under the right conditions, information derived from pharmacogenetics research is expected to improve the overall efficiency of drug development. It should also result in a better understanding of drug efficacy and adverse drug reactions (ADRs), much to the benefit of patients, health-care systems and drug developers. Under the right circumstances, pharmacogenetics might deliver economic growth and better health outcomes. The challenges to its adoption by drug developers and the health care sector are of interest, therefore, to governments.

Reducing risks, balancing benefits

Pharmacogenetics is poised to influence the health-care sector through its ability to reduce the risks associated with developing, funding and consuming pharmaceutical products. This offers potential benefits:

- Patients and health-care providers may face fewer risks because a better understanding of the variability of patient responses to drugs may lead to a quicker identification of the most efficacious therapies and reduce the risks of adverse events.

- Regulators and third-party payers may face less risk of adopting cost-ineffective drugs; less variation in patient response and fewer adverse events will result in savings in the health-care system as a whole.
- Drug developers may face lower financial risks: evidence suggests that the use of pharmacogenetic tools may reduce the probability and/or cost of product failure, the length of clinical development, and the need for large clinical trials and hence reduce the cost of drug development. Other things being equal, a greater number of effective treatments could enter the market for the same R&D expenditure.

While pharmacogenetics is clearly promising, its widespread adoption either by industry or the health-care sector is by no means certain. As of early 2009, there were just over a dozen pharmacogenetic-based products on the market, although some of the tests for pharmacodynamics can be used to determine the dosage of dozens of medications that are already on the market (see Table 1.1). For industry, investments in pharmacogenetics are plagued by uncertainty regarding future regulatory and reimbursement policies for pharmacogenetic drugs and associated genetic tests/diagnostics. In addition, the market size for pharmacogenetic drugs is unknown: pharmacogenetics typically targets patient population sub-groups and the size of these sub-groups may not be known until the drug development process is significantly advanced. Some industry stakeholders may perceive this "targeting approach" as potentially limiting their market size though some companies (*e.g.* large biotechnology firms) have already adopted a business model targeting smaller patient populations. Furthermore, in the clinical care setting, pharmacogenetics could make current models of care more complex and require physicians to be able to query and integrate new types of information from multiple sources in their treatment decisions.

Despite the complexity of this landscape, the use of pharmacogenetics is progressing rapidly. Its impacts are evident in three areas: *i)* basic research; *ii)* drug discovery and development; and *iii)* health-care management. This report will discuss what the impacts have been in these three sectors to date and what the challenges are to the broader uptake of pharmacogenetics throughout the innovation cycle (see Figure 1.1). The report specifically explores the socio-economic hurdles that could derail the anticipated benefits of pharmacogenetics. It also examines the assumption that pharmacogenetics can improve decision making and efficiency in the pharmaceutical R&D process and in clinical care. Finally, it examines what governments can do to create an "enabling" environment for the application of pharmacogenetics.

Table 1.1. Examples of personalised medicine applications

Product	Company	Technology/test type	Disease/application
HER2/neu tests	Several	Two types of test are available: immuno-histochemical tests measuring expression of the HER2/neu protein (phenotype) and FISH tests measuring amplification of the HER2/new gene (genotype)	Determine eligibility of breast cancer patients for treatment with Herceptin® (trastuzumab)
Trofile™ assay	Monogram Biosciences	Uses cultured cell lines to assess the interaction of the patient's HIV-1 strain with different cell-surface receptors (phenotype)	Determine eligibility of HIV patients for treatment with Selzentry® (maraviroc)
TPMT assays	Several	Two types of test are available, measuring the presence of TPMT gene variants (genotype) or the level of TPMT enzyme activity (phenotype)	Set dose of thiopurine drugs to maximise therapeutic efficacy while minimising bone marrow toxicity in diseases such as acute lymphocytic leukaemia, inflammatory bowel disease, and severe active rheumatoid arthritis
Invader® UGT1A1 assay	Third Wave Technologies	Uses PCR to measure presence of UGT1!1*28 gene variant (genotype)	Set dose of irinotecan in colorectal cancer patients to maximise therapeutic efficacy while minimising side effects of diarrhoea and reduced white blood cell count
AlloMap® test	Xdx	Uses PCR to measure expression of 20 genes, algorithm to convert results to quantitative composite score (multivariate genotype array)	Identify heart transplant patients at low risk of acute cellular rejection, may allow reduced use of biopsy for monitoring and/or more precise tailoring of immuno-suppressive regimen
Oncotype DX®	Genomic Health	Uses quantitative PCR to measure exoression of 21 genes, algorithm to convert results to quantitative composite score (multivariate genotype array)	

Source: Priorities for Personalised Medicine, Report of the President's Council of Advisors on Science and Technology, United States, 2008.

Figure 1.1.The innovation cycle

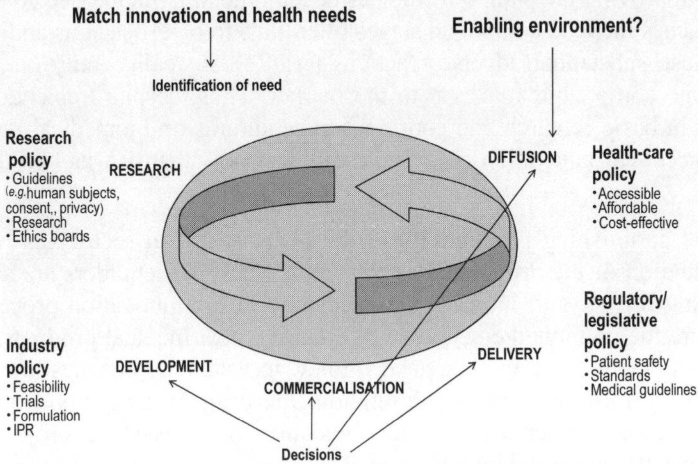

The pharmaceutical context

Pharmaceutical products are an essential component of the prevention and management of disease in all OECD member countries. Although pharmaceuticals are only a fraction of the medical care that individuals receive, they have proven to yield tremendous gains in care. They have helped treat many diseases which are causes of death and disability, and are one of the main reasons for the increased longevity of populations across the OECD. The continued vitality of the pharmaceutical innovation process is of great importance to OECD countries.

Over the last ten years, however, the drug discovery and development process has become increasingly lengthy and expensive, a problem compounded by the fact that there has been a decline in the number of new drugs discovered and introduced in medical practice globally. Developing a drug, from target identification to clinical use, currently takes an average of 12 years and costs an estimated USD 350 million to USD 1 billion. Much of the expenditure is linked to poor target identification and validation and to the failure (attrition) of lead compounds (often late in the development process). As industry addresses more difficult disease targets, attrition rates and costs have increased. For every 10 000 chemical compounds screened, approximately 100 candidate drugs are identified. Of these, only one reaches the lead compound stage, largely owing to the methodologies currently used in early-stage discovery and development, which are poor predictors of subsequent safety or efficacy. The main global regulatory bodies – the US Federal Drug Administration

(FDA) and the European Medecines Agency (EMEA) – have recognised that development of improved test methods early in the innovation cycle is a key to improving the path to medicines development. Even the one compound that reaches the lead compound stages often fails to be efficacious and may in fact cause substantial adverse reactions (ADRs). In reality, only one in ten candidate compounds makes it to the market. Thus, despite tremendous advances in basic research and enormous expenditures on biomedical research, pharmaceutical innovation is no more efficient now than it was two decades ago.

The diminishing productivity of the pharmaceutical sector has led to a re-evaluation of the drug discovery process. Many stakeholders are actively searching for ways to increase the efficiency of the innovation process and to ensure the continued emergence of effective new medical products. It has become evident that there is a need for new tools and approaches in order to make more informed choices throughout the drug discovery and development process, particularly in the translation of early-stage projects into medicines that meet public health needs.

Pharmacogenetics is a promising new approach. Unlike the current "blockbuster" business model employed by much of the pharmaceutical industry, which seeks to develop medicines that are "one size fits all", pharmacogenetics promises targeted therapy. Pharmacogenetics is based on the assumption that genetic heterogeneity is a significant source of variability in individuals' responses to drugs. Thus, accurate data on genetic differences within populations can facilitate rational drug discovery and delivery, and also minimise the incidence of ADRs during clinical trials and subsequent medical care. Pharmacogenetics essentially expands the toolkit for evaluating preclinical safety and efficacy as well as clinical utility. It can also, potentially, reshape how many medicines are discovered, developed and used. Two key benefits include its capacity to:

- reduce the costs of attrition of compounds during the drug development process.
- reduce the size, cost and duration of clinical trials.

Given the implications for pharmaceutical development costs, pharmacogenetics may have a significant impact on the productivity and competitiveness of pharmaceutical companies. Given its potential to reduce adverse drug reactions and improve overall patient responsiveness, it may also have a significant impact on health-care outcomes and reduce health-care costs substantially.[2]

The clinical context

Pharmacogenetics is already making significant contributions to the practice of evidence-based medicine, which is the deliberate use of the best available evidence when making decisions about patient care. When using evidence-based medicine, physicians weigh the relative advantages of different intervention choices and identify those that will be safest and most effective. With pharmacogenetics, an individual physician's clinical expertise is supplemented by external clinical evidence both from basic science and, most importantly, from patient-centred clinical research and practice (see Figure 1.1). It offers a set of methodologies that allow a critical appraisal of the accuracy and relevance of diagnostic tests, the power of predictive markers, and the safety and efficacy of therapeutic and preventative options.

A number of medicines for which pharmacogenetics-based tests exist are now on the market. These tests can differentiate patient groups and provide clinicians with guidance for prescribing decisions. In clinical practice, pharmacogenetic testing is currently applied for three reasons:

- to help identify responders and non-responders to a treatment.
- to aid in establishing appropriate dosages for responders.
- to identify susceptibility to ADRs and possibly exclude some patients from treatment.

Some examples of current uses of pharmacogenetics include: *i)* genetic testing to identify breast-cancer patients who are likely to benefit from treatment with Herceptin®; *ii)* identification of heart-failure patients as African-American for treatment with BiDil®, a medicine demonstrated to improve outcomes for individuals of African descent;[3] *iii)* genetic tests to identify patients who may react adversely to standard dosages of common medicines such as the anti-coagulant Warfarin®; *iv)* genetic tests for the identification of patients who may react adversely or not respond to tri-cyclic anti-depressants.

Despite early successes, recent reports note that pharmacogenetics has so far had only a weak impact on clinical practice and health care. A number of practical and technical challenges must be overcome before pharmacogenetics is introduced into mainstream health care. Pharmacogenetics provides probabilistic – rather than absolute – information that must be used with other medical information, since most diseases are complex and may involve more than one gene with variable penetrance. The behaviour of these genes also depends on environmental factors such as age, sex and diet. In essence, pharmacogenetics aims to increase the probability of a correct decision in drug prescription.

Thus, pharmacogenetics presents something of a double-edged sword for health professionals and health systems. More effective prescribing and better health outcomes are on offer, but system complexities will increase compared to current models of care. Physicians will need tools to capture and integrate a number of environmental factors such as co-medication and co-morbidity. They will also want to be able to compare new approaches to existing models of care, and to confirm therapy across a number of diagnostic biomarkers and platforms. To do so, they will need to query significant amounts of patient data to gain insight into disease prevention, diagnosis and treatment. In short, physicians will need access to effective data query and data mining tools. For all these reasons, there is debate about whether pharmacogenetics will be as clinically useful or economically beneficial for the health-care system as might be hoped.

The preceding factors emphasise the need for systematic evaluation of the costs and benefits associated with the introduction of pharmacogenetic testing to guide drug prescribing. Such an evaluation will need to consider the perspectives of all relevant stakeholders and will need to be carried out within the context of the entire cycle of innovation (see Figure 1.1).

The regulatory and ethical contexts

The use of pharmacogenetic information also raises fundamental as well as practical questions for regulatory systems as well as issues of ethics and equity. Pharmacogenetics is driving some re-evaluation of regulatory processes, including safety assessment, post-authorisation use and safety-monitoring procedures. Such evaluation includes considerations of whether existing medicines need to have their patterns of use or dose regimes changed.

Pharmacogenetic applications are also raising important social and ethical issues. Major areas of concern include:

- the storage of DNA samples and genetic data.
- ethical issues related to the reshaping of clinical trials based on genotypes of patient groups, including informed consent for such trials.
- the equity of health-care delivery, particularly in terms of possible exclusion from clinical care that involves potentially life-saving drugs.

There is a need to develop policy frameworks that can accommodate the complexities of ethical views, as well as the cultural and population diversity of countries. A wide range of stakeholders – patients, regulators, providers, insurers, the research community and the pharmaceutical industries – have a vested interest in understanding how advances in pharmacogenetics will affect both the discovery, development, evaluation and use of new and existing medicines, and the related regulatory frameworks.

In order to seize the opportunities – and meet the challenges – offered by pharmacogenetics, sectors will have to go through a period of joint evolution. The degree to which there is co-operation between the public and private sectors, as well as across national borders, may be fundamental to determining the extent to which society as a whole benefits from the promise of genomics to deliver better health. It may also determine the extent to which society realises a successful return on public and private investment in the Human Genome Project and similar ventures.

The framework for the OECD pharmacogenetics and health innovation project

In October 2005, the OECD held a workshop in Rome entitled "An International Perspective on Pharmacogenetics: the Intersections between Innovation, Regulation and Health Delivery". It was co-sponsored by the governments of Italy, Canada and Australia, and supported by a special project fund of the Business and Industry Advisory Committee (BIAC). The workshop's main objectives were to:

- provide an overview of advances in pharmacogenetics research internationally.
- analyse anticipated impacts of pharmacogenetics on innovation, health delivery, and health-care systems.
- identify regulatory issues that arise in translating the outputs of pharmacogenetics research into products for targeted therapies and diagnostics.
- identify possible initiatives and strategies relevant to the development and implementation of pharmacogenetics, and to the improvement of the delivery of health innovation and public health across OECD countries.

This report draws on the discussions that took place at the Rome workshop. However, it goes somewhat beyond the topics addressed at the workshop and is not intended to be read as the proceedings of that event.

The analysis carried out in this report is based on a "systemic' or holistic framework of the health innovation cycle (see Figure 1.1). Health innovation is an interactive and distributed process which involves a number of phases: *i)* identification of need; *ii)* research and development; *iii)* commercialisation; *iv)* delivery; and *v)* diffusion. Rather than see these stages are separate and linear, the framework used views the stages as circular and iterative (in fact Figure 1.1 truncates the phases into four – discovery, development, delivery and diffusion). This way of envisaging the innovation process is radically different from the closed, sequential, linear perspective of innovation (Figure 1.2) that policy so often assumes.

The bio-medical research community, the bio-pharmaceutical industry, and the medical service economy are interlinked. Health innovation is closely connected to the provision, uptake and use of new treatments: feedback from purchasers, providers and patients is essential in shaping the innovation process. Feedback mechanisms are built in throughout this innovation cycle, and are the source of modifications that improve individual products and enhance innovative capacity as a whole.

Figure 1.2. The linear model of innovation

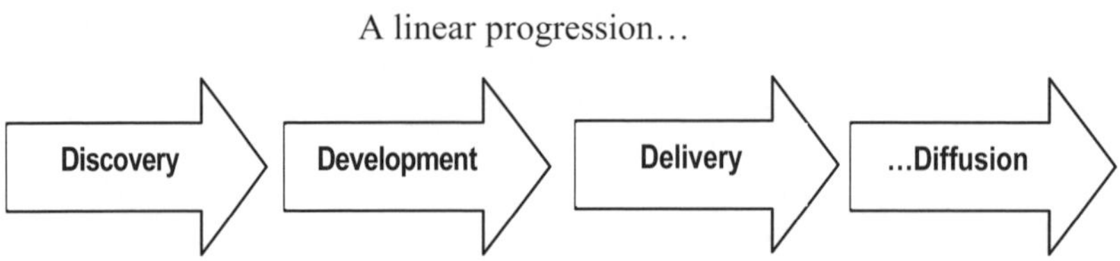

This cyclical model arguably more accurately represents the process of health innovation than the traditional linear model for some time. However, the advent of pharmacogenetics and the concomitant possibility explicitly to link individuals' genetics with the action of a medicine and with patient outcomes create the opportunities and arguably the necessity for actual patient outcome data to be compared with predicted outcome data so as to improve rational drug design. Innovation can no longer be regarded as a unidirectional linear process.

Opportunities for new health technologies cannot and must not be separated from longer-term opportunities in health-care delivery. More specifically, the widespread adoption and use of pharmacogenetics to deliver better public health will not happen on the basis of scientific advances alone: it requires the right set of policies to be in place both for the producers in the health-care industry to invest in the development of such therapies and for the users – patients, doctors and reimbursement systems – to recognise a clear clinical benefit.

Structure of the report

This report on the opportunities and challenges posed by pharmacogenetics identifies needs of and incentives for the many stakeholders that participate in the health innovation cycle. Its aim is to provide guidance for policy makers.

Chapter 2 reviews the supporting technologies and infrastructures that will accompany the use of pharmacogenetics. It argues for a better integration of new scientific and technological advances in the process of target identification and validation. A major section of the chapter focuses on biomarkers, which are emerging as one of the most effective ways to address the efficiency problems in drug discovery, and offers an overview of the potential applications of genetics-based biomarkers. The chapter also considers the information systems that must be in place to support the clinical application of pharmacogenetics. These systems must meet the demand for easily accessible, accurate and timely information which can be collected, collated and exchanged among health-care professionals, for analysis and decision support at point of care.

Chapter 3 examines the application of pharmacogenetics to drug development. It explains why there has been a recent slowdown in the discovery and development of new medicines and how pharmacogenetics may help address this problem. It discusses how two of the principal applications of genetic biomarkers will affect drug development. Genetic biomarkers can demonstrate drug efficacy and explain adverse events at a molecular and genetic level; and identify the patient groups that are most likely to benefit from a particular medicine. Using genetic screening to identify potential responder populations prior to enrolment in clinical trials could enable firms to demonstrate drug efficacy in smaller sets of subjects. Such an approach can, in principle, reduce the costs of clinical trials for prospective medicines while increasing the probability that such potential products might reach the market.

Chapter 4 discusses the challenges posed by pharmacogenetics to established "big pharma" business models. The pharmaceutical and the diagnostics and device industries face very different sets of incentives for integrating pharmacogenetic approaches in their discovery and development activities. In the pharmaceutical industry, pharmacogenetics puts pressure on the blockbuster business model. In the devices and diagnostics industry, pharmacogenetics creates uncertainties about how to capture value when co-developing and co-marketing assays and therapeutics. The chapter discusses the viability of current business models and the financial incentives of different types of firms.

Chapter 5 addresses pharmacogenetics end users: physicians, patients and health-care systems. Two factors in particular will influence the development of pharmacogenetics: the extent of recognition by decision makers and the medical community of the need for safer and more effective drugs; and the ability of pharmacogenetics to identify which interventions will or will not work. The chapter reviews current evidence on physicians' attitudes to pharmacogenetics. It also covers patients' perspectives and possible ethical and equity issues, particularly those concerning developing countries. It closes with consideration of the opportunities and challenges for incorporating pharmacogenetics in clinical care as well as potential organisational and system-wide hurdles. Although it is likely that pharmacogenetics will only affect specific therapeutic areas (*e.g.* oncology) and drugs in the short term, it is reasonable to assume that the technology will eventually be applied to both common and complex diseases. This will necessarily create both technical and educational challenges for health professionals. Successful integration of pharmacogenetics will depend upon appropriate infrastructure and trained personnel.

Finally, Chapter 6 looks at the role of regulators and reimbursement systems in facilitating the uptake of pharmacogenetics by health-care systems. Regulators have a vital role to play in enabling the application of pharmacogenetics for both new and established drugs. Coverage policies will also be a strong factor in the rate of diffusion of pharmacogenetics. In particular, reimbursement policies and cost-containment strategies for drug and test use policies and clinical guidelines will be very important. The chapter concludes with a discussion on the challenges for payers and the trade-offs they may have to make.

Notes

1. As well as such ministers from China, Israel, Russia and South Africa.
2. Reducing adverse drug reactions has several manifestations. It may mean reducing the number of patients that feel unwell or ill at ease when taking medication and thus discontinue therapy, foregoing the potential benefit. It may also mean avoiding toxicity events that require palliative care. And it may mean avoiding a situation in which there is no adverse event yet no positive therapeutic impact.
3. Such identification is carried out by the patients themselves.

Chapter 2

The Supporting Infrastructures for Pharmacogenetics

This chapter reviews the technologies and infrastructures that will accompany the use of pharmacogenetics and argues for a better integration of new scientific and technological advances in target identification and validation. A major section focuses on biomarkers and offers an overview of the potential applications of genetics-based biomarkers. The chapter also considers the information systems that must be in place to support the clinical application of pharmacogenetics.

The significant advances in genomics-based sciences of the last few years have spawned two rapidly evolving disciplines: pharmacogenomics and pharmacogenetics. Both are expected to have a strong impact on the future of drug development and delivery. They are providing a new body of knowledge about diseases and about the compounds that might provide effective treatment against them.

Of the two disciplines, pharmacogenetics is the main focus of this report. It is a scientific discipline developed by the biomedical and clinical research communities which recognises that because of genetic differences, not all individuals who exhibit a particular disease will experience the same response to drug therapies. As a result, knowledge gained from genetic research can be used to direct medical care, particularly to target drug therapy to specific patient groups. It achieves this, in part, by identifying biomarkers, indicators that mark the presence of potential gene-drug interactions and measure response to therapeutic activity.

However, the speed of progress that can be achieved with pharmacogenetics will depend on advances in other scientific fields and technologies. Fields and technologies of particular importance for pharmacogenetics include:

- systems biology,
- rapid mapping of single nucleotide polymorphisms (SNPs),
- high-throughput assays (*e.g.* microarrays),
- human genetic research databases,
- bioinformatics and interoperability of datasets,
- statistical analysis.

For pharmacogenetics to deliver on its promise, correlations between genomic variation and variation in drug-related phenotypes (at different levels of detail – molecular, cellular, organ and organism) must be identified and verified. Large population-based data sets, such as human biobanks and genetic research databases (HBGRDs), can serve as resources for storing and searching libraries of genetic material and other clinical information. This will require linking biological and clinical information and the various data sources will have to be made interoperable. As the data will most likely be owned by different groups, new mechanisms may be needed to facilitate the sharing of data, information and research results.

What are pharmacogenetics and pharmacogenomics?

There is sometimes a lack of precision over the use of the terms "pharmacogenomics" and "pharmacogenetics" which may inappropriately be used interchangeably. Furthermore, where they are defined, the definitions may vary. For the purposes of this report, which is predominantly concerned with pharmacogenetics, its objective is considered to be to improve the safety and efficacy of new and existing medicines. In contrast, the aim of pharmacogenomics is broader and more closely linked to understanding genetic systems, including their interactions and influences.

The term pharmacogenetics was first used in 1959 to describe a scientific discipline that sought to understand how the genetic make-up of individuals influenced their responses to medicines. In this report, pharmacogenetics is defined as the study of the effects of variations in DNA sequence (genetic differences) on drug response, in terms of both the metabolism (pharmacokinetics) and the action (pharmacodynamics) of the drug delivered. In simple terms, the discipline aims to identify the best medicine for a specific disease when the disease occurs in a patient population with a particular genotype.

The pursuit of pharmacogenetics is important because it is estimated that genetic factors account for 20-95% of the observed responses to drug therapies, depending on the drug and on the genotype of the target population. For seven out of 14 major drug classes, it has been estimated that 50% or less of patients respond. Other factors that influence an individual's response to a particular drug include age and the simultaneous presence of other diseases (co-morbidity) or of other substances, either through diet or co-medication.

Pharmacogenomics, on the other hand, aims systematically to assess how interacting systems of genes may affect disease susceptibility, pharmacological function, drug disposition and therapeutic drug response. One of its primary aims is to identify genetic markers that may assist in the diagnosis, staging, and classification of disease within the context of drug responses. It also seeks to optimise the identification of those drugs in the discovery pipeline that induce the most desirable pharmacological responses. Pharmacogenomics is expected to have a major impact on the drug discovery procedures currently used by the major pharmaceutical companies.

Pharmacogenetics is essentially a subset of pharmacogenomics: it can have an impact on both new and existing medicines and is expected to have a major impact on the translation of early-stage projects into medicines.

Technological developments

Several recent advances in genetic research have made the application of pharmacogenetics both more feasible and more important. In-depth study of the human genome has shed new light on the sources of individual genetic variations and their effects. The most common variations are substitutions of a single nucleotide for another at a given location in the base sequence of a gene. These are known as single nucleotide polymorphisms (SNPs). Combinations of SNPs which aggregate together are known as "haplotypes"; they are useful genetic markers for indicating a potential drug response in patients.

Three key projects – the sequencing of the human genome, the development of the SNP Consortium, and the publication of the HapMap – have collectively resulted in a body of data that is accelerating the study of genetic aspects of drug response (see Box 2.1). In support of these publicly available data sets, new technological platforms and tools were developed. Gene expression arrays, for one, provide genome-wide analysis of gene expression. Other tools assay the response of cells and organisms to different circumstances or challenges at a genetic level: 2-D differential gel electrophoresis, used for the separation and identification of proteins; sophisticated mutation assays (such as transposon-mediated differential hybridisation); and use of small interfering RNA (siRNA) to study gene function.

Box 2.1. Collaborative mechanisms, knowledge networks and consortia for pharmacogenetics

Numerous collaborative mechanisms, knowledge networks and consortia have emerged and are addressing specific aspects of pharmacogenetics.

The **International HapMap Project** (*www.hapmap.org*) is a collaboration between scientists and funding agencies from Canada, China, Japan, Nigeria, the United Kingdom and the United States. The intention is to compare the genetic sequences of different individuals in order to identify chromosomal regions in which there are shared genetic variants. This information will be catalogued and made available in the public domain, so that it can be used by other researchers to discover the genetic variants involved in disease and individual response to therapeutic agents.

The **Japan Pharmacogenomics Consortium** (JPGC; *www.jpgc.org/e/index.htm*) was established by 11 Japanese companies in 2003 to promote the development of infrastructure and national standardisation for pharmacogenomics-based clinical trials in Japan, including post-marketing trials (Phase IV clinical trials).

…/…

Box 2.1. Collaborative mechanisms, knowledge networks and consortia for pharmacogenetics *(continued)*

The **Pharmacogenetics for Every Nation Initiative** (PgENI; *http://pgeni.im.wustl.edu*) aims to develop a pharmacogenetic approach using self-defined ethnicity as the clinical marker. This would help in identifying populations throughout the world that are at elevated risk of toxicity or treatment failure when using the WHO essential medicines list. The project also aims to provide information that will facilitate the selection, from WHO-recommended therapies, of treatment regimes for a given population. The initiative is concerned that many medicines are less effective for, or more dangerous to, populations that are of non-European origin (Kiman *et al.*, 2004).

The **Pharmacogenetics and Pharmacogenomics Knowledge Base** (PharmGKB) is an online resource *(www.pharmgkb.org)* that provides information on the relationships among drugs, diseases and genes, including their variations and related products. It has a dual mission: *i)* to catalyse pharmacogenomics research by developing, implementing and disseminating a public genotype-phenotype resource focused on pharmacogenetics and pharmacogenomics; and *ii)* to perform high-quality research in support of this goal, and to catalyse scientific discovery in both pharmacogenetics/pharmacogenomics and in biomedical informatics. Thus, in the short term, the knowledge base will facilitate basic research. In the long term, it will influence how medicine is delivered.

The **Pharmacogenetics Research Network** (PGRN; *www.nigms.nih.gov/Initiatives/PGRN*) is a US-based network of scientists focused on understanding how genes affect a person's response to medicines. Its long-term goal is to make information available to doctors, thereby helping to ensure that the right dose of the right medicine is given – the first time – to everyone. The PGRN comprises 12 independently funded interactive research groups, each of which specialises in a different area of pharmacogenetics (see *www.pharmgkb.org/network/members.jsp* for a description of the research groups). The PGRN is conducting studies of human gene variations relevant to pharmacokinetics (drug disposition) and pharmacodynamics (drug action), as well as the relationship of such variation to drug-response phenotypes. The resulting data is deposited into the knowledge base, PharmGKB *(www.pharmgkb.org)*.

The **Public Population Project in Genomics** (P3G; *www.p3gconsortium.org*) is a not-for-profit international consortium to promote collaboration between researchers in the field of population genomics. It was launched to provide the international population genomics community with the resources, tools and know-how needed to facilitate data management for improved methods of knowledge transfer and sharing. Its main objective is to create an open public and accessible knowledge database.

The **SNP Consortium** (TSC; *http://snp.cshl.org*) was established in 1999 as a collaboration of several companies and institutions to produce a public resource of single nucleotide polymorphisms (SNPs) in the human genome. The initial goal to discover 300 000 SNPs in two years was far exceeded; by 2001, information on more than 1.4 million SNPs had been released into the public domain. Now that the SNP discovery phase of the project is essentially complete, the emphasis has shifted to studying SNPs in populations.

The **Structural Genomics Consortium** (SGC; *www.sgc.ox.ac.uk*) is a not-for-profit partnership between the Wellcome Trust and other funders including pharmaceutical companies, three Canadian funding bodies and a Swedish consortium. Its goal is to determine the three-dimensional structure of proteins of medical relevance, and to place these in the public domain. The SGC is targeting 375 proteins for the first four years.

Advancing pharmacogenetics through knowledge networks and open innovation

The biomedical research community has taken the opportunity offered by the major data sets and new technological developments to investigate the genetic factors governing the response of both individuals and population groups to diseases and drugs. A large body of information is being created which improves our understanding of the potential of pharmacogenetics. Three new avenues of research include:

- studies to identify associations between SNPs and other genotype variations with a number of diseases;
- the use of genetic markers to predict response to medicines;
- the examination of impacts of ethnicity on drug response.

Health care is unlikely to benefit from increased data collection alone. Efforts need to be made so that the data generated, and the research that produced them, have genuine clinical utility (*i.e.* they are not just of biological interest) and are relevant to patients. By way of example, a pharmacogenetic test to ascertain the need to modify a dosing regimen for only 5% of the potential recipient population might not be clinically useful, unless that dosing modification prevented serious/severe adverse reactions, whereas there is more likelihood that a test covering 100% of that same population would have utility.

In fact, the real added value of pharmacogenetics will be realised when various data sets have been effectively linked so that initial hypotheses about possible correlations between genomic variations and therapy-related effects can be verified and validated. To do this, genetic and genomic data will need to be integrated with phenotypic data – a significant challenge in that these data are defined at levels ranging from the molecular to the clinical. Moreover, clinical studies involving a large number of patients from a variety of population groups are necessary to gather evidence of clinical utility. The large scale of such patient cohort studies makes them too expensive to be undertaken by conventional research groups: collaborative approaches to sample gathering and storage will be necessary in many cases.

This need for collaboration draws attention to the ways in which information will be collected, stored and used. As with the human genome and other collaborative projects (see examples in Box 2.1), collaborative approaches will need to include agreements relating to the availability of raw data and the sharing of data. This will, in turn, depend on knowledge and intellectual property management systems that support open innovation platforms for pharmacogenetics.

Such collaborative approaches are necessary if society is to reap the full benefits of pharmacogenetics. But collaboration introduces complexities, and ways (both technical and policy tools) will need to be found to manage issues such as:

- technical challenges related to sharing and interoperability of data;
- privacy and security of data, from the technical, ethical, legal and social perspectives;
- quality assurance of testing and data handling as well as reporting to patients;
- potential that data could be misused to genetically discriminate against individuals or populations;
- appropriate respect for applicable intellectual property provisions and downstream behaviour that favours collaboration;
- the risk of anti-competitive behaviour that breaches anti-trust rules.

Various safeguards of course are already in place to manage such issues – not least at the level of the OECD itself.[1]

Human genetic research databases: privacy and security issues

Certainly, successful exploitation of pharmacogenetics research will require the creation and utilisation of large-scale human biobanks and genetic databases (HBGRDs), which raises many of the issues touched upon above. The OECD recently published a report based on a conference, the *Creation and Governance of Human Genetic Research Databases* (OECD, 2006a). The report concludes that existing legal frameworks for consent and privacy may create obstacles to the successful development of HBGRDs which will need to be navigated. For example, in many countries it is necessary to satisfy a number of national and regional privacy laws and policies before a large-scale HBGRD can operate. These may include national policies and guidelines related to research ethics, professional standards, privacy legislation and human tissue laws. In some jurisdictions, judge-made or common laws on consent must also be considered (Australia Law Reform Commission, 2003). To complicate matters further, in a number of countries (*e.g.* Canada and Australia) relevant consent and privacy laws vary widely between regional jurisdictions (*i.e.* provinces and states).

Security issues are also of critical importance. Within a health-care setting, decision making revolves around the aggregation and analysis of individual patient data which, although coded, must remain identifiable to enable clinical care. Current legal systems in most OECD countries provide guidelines for maintaining patient data confidentiality. However, the sheer volume of data, as well as the potentially large number of end users in a clinical setting, calls for highly secure systems that incorporate sophisticated tools for authentication of authorised users and effective measures to monitor access to patient data. The challenge is to build systems that have the flexibility to accommodate new data inputs as they arise and are capable of providing output to specialised tools to support clinical decisions.

As noted above, new and existing tools to manage the successful use of pharmacogenetics data will need to be developed and aligned. Current OECD work towards developing Council Guidelines on Human Biobanks and Genetic Research Databases will make a contribution towards this goal.

Target identification and validation

Current practices and their shortcomings

A particularly complex aspect of drug development is the identification and validation of drug targets. Estimates suggest that up to USD 100 million (roughly 10% or more) of the cost of each drug development programme is lost at the interface between preclinical and clinical research. As a rule of thumb, for every 100 000 chemical compounds initially screened and found to have some potential, approximately 100 (or just 0.1%) are identified for further investigations. Of these 100, typically only one becomes a new candidate for drug development, *i.e.* makes it to the "lead compound" stage. Many of these lead compounds – by some estimates 40-60% –fail to meet what are known as the ADMET properties (for a definition see Box 2.2). In the 1990s, poor biopharmaceutical properties (*e.g.* drug absorption, bio-availability) accounted for approximately two-fifths of lead compound failures.[2] By 2000, these were much less frequently a cause of attrition as they are identified earlier in the drug development process, while safety and toxicity issues now accounted for more than 30% of the drug failures.[3] Such attrition rates are crippling for the pharmaceutical industry. Hence, improving the effectiveness of target identification and validation is central to the success of pharmaceutical R&D.

Box 2.2. The ADMET properties

In OECD countries regulations require that the pharmacokinetics of new candidates for drug development are well understood – in particular the ADME-Tox properties describing the disposition of a pharmaceutical compound within an organism. The properties that help understand how a pharmaceutical reaches its intended target and passes through the body are:

Absorption: The ability of a drug to pass from the digestive tract to the bloodstream.

Distribution: The amount of a drug that various tissues extract from the blood.

Metabolism: The chemical changes the body makes to the drug.

Elimination: How the body expels the drug.

Toxicology: The adverse effects of the drug on the body, ranging from minor symptoms (*e.g.* nausea) to death.

For more information see R.A. Lipper (1999), *Modern Drug Discovery*, 2(1), ACS Publication, Washington, DC, pp. 55-60.

Traditionally, decision making about compounds early in the development process has been based on methods such as random screening, laboratory tests, computer models and animals studies. These efforts are often frustrated by a paucity of experimental data which define the structure and properties of the biological target. If they were available, such data could be used to predict the likely clinical success or failure of the selected compounds. This lack of data can cause compounds to fail or significantly slow down during the development process for a number of reasons.

This explains why, over the past five to ten years, the pharmaceutical industry has primarily focused its drug portfolios on a few hundred relatively well-characterised therapeutic targets. The net result has been the introduction into the health-care system of a larger number of non-innovative "me too" drugs, or variants of existing medications, rather than entirely novel or "first-in-class" medicines. From 1989 to 2000, the US Food and Drug Administration (US FDA) approved more than 1 000 drugs. Only 361 of these were "new" enough to be classified as new molecular entities (NMEs); the others contained active ingredients already approved. Moreover, recent analysis of 400 established drug targets sought to identify the families to which they belong; it showed that only 130 gene families are represented (Hopkins and Groom, 2002). Such analyses suggest that this source of targets has been exhausted and that new paradigms for target identification and validation are needed.

Box 2.3. Biomarkers: a potential breakthrough in drug development

Histological markers to identify responders

Example: Standard imaging studies are limited in their ability to act as outcome measures for patients with metastatic prostate cancer. Fluorodeoxyglucose positron emission tomography (FDG-PET) scans can serve as an outcome measure for patients treated with chemotherapy. Fluorodeoxyglucose-Positron Emission Tomography (FDG PET) imaging allows the detection of viable tumour cells independent of morphology. Deoxyglucose labelled with the positron emitter fluorine is a glucose analogue that is transported into cells like glucose. Unlike glucose it is trapped in cells in a phosphorylated form after uptake and further metabolism does not occur. Since cancer cells have a higher glycolytic rate than normal cells, the tracer concentrates in neoplastic tissues and allows their detection.

Physiological markers used to predict at-risk patients

Example: Prostate specific antigen (PSA) is produced by cells in the prostate. Under normal circumstances, the prostate secretes PSA into semen (PSA prevents coagulation of semen, thereby assisting in reproduction). Small amounts of PSA naturally leak out into the bloodstream as well. When prostate cancer is present, the prostate ducts that normally secrete PSA into the urethra become clogged; more PSA leaks into the bloodstream. Elevated PSA levels in the bloodstream do not confirm the presence of cancer, but they do indicate the need for further investigation.

Pathological distinctions differentiate diseases

Example: The human epidermal growth factor receptor-2 (HER2/*neu*) is a well-characterised pathological biomarker in the biology of breast cancer which has had immediate impact on clinical medicine. Research has shown that HER2-positive breast cancer is a more aggressive disease with a greater likelihood of recurrence, a poorer prognosis and a decreased chance of survival compared with HER2-negative breast cancer. Identification of those breast cancers that show HER2-positive characteristics distinguishes a group of cancers with a less good prognosis and offers opportunities for targeted treatment.

Chemical analysis guides dosing

Example: Monitoring of drug levels within the bloodstream provides chemical information that guides the clinician in making dose-level decisions. It can also help identify patients who may be poor or ultra-rapid metabolisers. Biomarker monitoring can also indicate the potential development of side effects that would influence prescribing decisions, as, for example, with the use of acetretin in the case of liver damage.

Clinical markers predict disease process

Example: Blood pressure (BP) is monitored as a measure of patient health in a wide variety of contexts. Significant changes in blood pressure can be an indicator of disease (as in the development of cardiovascular disease) or of the developments of unwanted side effects (as with oral contraceptives). Reduction in blood pressure can indicate the effectiveness of a course of treatment or other action taken by the clinician or patient.

Source: Lesko (2005), presentation at OECD workshop "An International Perspective on Pharmacogenetics: The Intersections between Innovation, Regulation and Health Delivery", Rome, Italy, 17-19 October, *www.oecd.org/sti/biotechnology/pharmacogenetics*.

Prospects and limitations of biomarkers

In recent years, key advances have greatly expanded the universe of chemical compounds from which to pick drug candidates for development. To a large degree, the discovery of new compounds can be attributed to the coupling of genomics and proteomics with new technologies and methods. These new technologies and methods include high-throughput screening (the process by which large numbers of compounds are tested in parallel for binding activity or biological activity against target molecules), combinatorial chemistry and structure-based design (Greer *et al.*, 1994; Bohacek *et al.,* 1996).

Systems biology, including whole-pathway approaches, is also emerging as a fundamentally new strategy for selecting drug candidates. With systems biology, researchers combine data about genes, proteins and metabolites to generate a comprehensive picture of the connections between the various components in a biological system. In the context of drug discovery, the data can be used in computer models and simulations to identify the pathways involved in a particular disease. This may lead to new therapeutic targets or may help to determine whether – and how – a compound is following the intended pathway.

A major goal of many systems biology projects is to identify characteristics that can act – alone or in combination – as indicators of the behaviour of the system. These are commonly referred to as "biomarkers". Biomarkers, in and of themselves, are not novel. They are specific physical traits that can be used to indicate the presence or measure the severity of a disease or condition – or to measure the effect of a therapeutic activity. They can include anatomical, physiological, biochemical or molecular parameters (see examples in Box 2.3). They can help identify points at which "healthy" systems begin to deteriorate or at which therapies elicit the desired response to improve health – or fail to do so. Therefore, they can be used to assess the efficacy of a drug or to expose unexpected toxicity, the main causes of failure in late-stage drug development.

The lack of useful biomarkers is arguably the main cause of current inefficiencies in translation from the preclinical to clinical phases in drug development. To be useful, a biomarker needs to exhibit three features. First, it must be measurable. Second, it must be causally involved and have a central role in the emergence, severity, progression or reversal of a disease (so that dissociation from clinical outcome is unlikely). Third, a biomarker should be easy to measure accurately and reproducibly.

Biomarkers can act as surrogates for outcome measures or clinical endpoints. A clinical endpoint is a characteristic or variable that reflects how a patient feels, functions or survives. Surrogate endpoints, as used in the scientific and regulatory communities, are findings or measurements that may be used in clinical trials to evaluate the safety or effectiveness of a medical therapy for treating disease. They could serve as an alternative to the most common clinical endpoints of morbidity and mortality. A few biomarkers are so closely linked to the clinical outcome that regulatory authorities consider them valid primary variables in clinical trials and adequate substitutes for clinical endpoints during the marketing authorisation process. Surrogates are useful in that it generally takes considerably less time and fewer subjects to demonstrate a significant effect of a drug.

Biomarkers are therefore one of the most effective means of improving the efficiency of the drug discovery process. Their use will have an important impact on the efficiency and productivity of translational research and clinical practice.

However, significant technical limitations must still be overcome. As discussed, new technologies are being applied to identify and validate pharmacogenetic biomarkers. For example, different molecular profiling methods – such as DNA microarrays and proteomics – are being used to detect unique "fingerprints" of molecular changes due to disease, drug treatment or toxicity. Each fingerprint reflects a cumulative response of complex molecular interactions. If these interactions can be significantly correlated to an end point, the molecular fingerprint may qualify as a predictive biomarker. However, much of the data is currently produced in various formats and units, and is normalised and analysed in different ways. For validation studies to be carried out, appropriate frameworks, systems and methodologies must be put in place. Research and eventual applications also require the development of standards for encoding and processing information – in other words, for ensuring the interoperability of distinct data sources.

Clearly, there is a need to co-ordinate research and reporting among biomarker-development laboratories, biomarker-validation laboratories, clinical repositories and population-screening programmes. It is hoped that co-ordination of research efforts will facilitate collaboration and promote efficiency and rigour in biomarker research. Standards for encoding and processing information – in other words, for ensuring the interoperability of distinct data sources – will be critical in facilitating this research and related applications.

Conclusions

Pharmacogenetics is expected to have a major impact on both basic and applied research, particularly the translation of early-stage projects into medicines. Its main potential lies in its ability to identify biomarkers or indicators that mark the presence of potential gene-drug interactions or measure response to therapeutic activity.

Several recent advances in genetic research and new technological platforms have made pharmacogenetics more feasible and more important. The biomedical research community is developing a large body of information which improves current understanding of the response of individuals and subpopulations to diseases and drugs at a genome-wide scale. This research is beginning to provide biomarkers that are useful to researchers, including in the private sector, for improving the drug development process. The impact of biomarkers will be felt throughout the drug development process and into clinical practice. Biomarkers will be used to predict drug response, to differentiate and rank lead compounds, to accelerate preclinical and clinical studies, and to identify individuals most likely to benefit from therapy. Biomarkers can help to reduce the size and cost of clinical trials, thereby bringing useful medicines to patients more quickly and rapidly and decreasing the economic burden of clinical development.

However, the broader uptake of pharmacogenetics faces a number of hurdles. One hallmark of pharmacogenetic research studies is the large amount of genetic data that must be accumulated and integrated for high-resolution drug-response genotyping and subsequent phenotype profiling. Collecting the data on drug response variability that are necessary in order to make genotyping clinically predictive can be a complex and expensive undertaking. In some cases it might require information on a few polymorphisms or genes; in others it might require complex studies involving a relatively large number of genes or a genomic-based approach.

Infrastructures to enable such data gathering, analysis and validation might include:

- multidisciplinary international networks that can increase the efficiency of research;
- public-private partnerships to carry out prospective studies;
- agreements relating to the availability of raw data and the sharing of data;

- systems to manage knowledge and intellectual property so as to support open innovation platforms for pharmacogenetics;
- the creation and utilisation of large-scale human biobanks and genetic databases (HBGRDs).

The aggregation and use of large amounts of human genetic data will present their own challenges, in terms both of development and alignment of policy and of technical tools.

In addition, to bridge the gap between preclinical and clinical development, two other hurdles must be addressed. Each of these is taken up in further chapters: *i)* the weak economic incentives for the private sector to develop biomarkers; and *ii)* the lack of harmonisation across jurisdictions for the validation of biomarkers and related regulatory issues. At present, many different technologies are used to identify and validate biomarkers. Hence, much of the data is produced in various formats and units, and is normalised and analysed in different ways. Appropriate frameworks, systems and methodologies must be established to carry out the required validation studies.

Notes

1. OECD Council Recommendations providing the *OECD Guidelines for Licensing of Genetic Inventions* (2006) and the *OECD Guidelines for Quality Assurance in Molecular Genetic Testing* (2007).
2. See Lipper (1999).
3. See Smith and O'Donnell (2006).

References

Australian Law Reform Commission (ALRC) (2003), *Essentially Yours: The Protection of Human Genetic Information in Australia*, Australian Government, Report 96, March.

Bohacek, R.S., C. McMartin and W.C. Guida (1996), "The Art and Practice of Structure-based Drug Design: A Molecular Modeling Perspective", *Medicinal Research Reviews*, no. 16, pp. 3-50.

Greer, J., J.W. Erickson, J.J. Baldwin and M.D. Varney (1994), "Application of the Three-dimensional Structure of Protein Target Molecules in Structure-based Drug Design", *Journal of Medicinal Chemistry,* no. 37, pp. 1035–1054.

Hopkins, A.L. and C.R. Groom (2002), "The Druggable Genome", *Nature Reviews* No.1, pp. 727-730.

Kiman, K., J.A. Johnson and H. Derendorf (2004), "Differences in Drug Pharmacokinetics Between East Asians and Caucasians and the Role of Genetic Polymorphisms", *Journal of Clinical Pharmacology*, No. 44, pp. 1083-1105.

Lesko, L. (2005), "An International Perspective on Pharmacogenetics", presentation at the OECD workshop "An International Perspective on Pharmacogenetics: The Intersections between Innovation, Regulation and Health Delivery", Rome, Italy, 17-19 October, *www.oecd.org/sti/biotechnology/pharmacogenetics.*

Lipper, R.A. (1999), *Modern Drug Discovery*, 2(1), ACS Publication, Washington, DC.

OECD (2006a), *Creation and Governance of Human Genetic Research Databases*, OECD, Paris.

OECD (2006b), *OECD Guidelines for Licensing of Genetic Inventions*, OECD, Paris.

OECD (2007), *OECD Guidelines for Quality Assurance in Molecular Genetic Testing*, OECD, Paris.

Smith, C.G. and J. O'Donnell (2006), *The Process of New Drug Discovery and Development,* CRC Press, Boca Raton, FL.

Chapter 3

Pharmacogenetics and Drug Development

This chapter examines the application of pharmacogenetics to drug development. It explains why there has been a recent slowdown in discovery and development of new medicines and how pharmacogenetics may help address this problem. It discusses how application of genetic biomarkers will affect drug development. Such an approach can, in principle, reduce the costs of clinical trials for prospective medicines while increasing the probability that potential products might reach the market.

During the past decade, there has been a decline in the productivity of drug development. Despite the tremendous recent success of basic research and the significant expenditures by both public and private sectors to support biomedical research, the pharmaceutical innovation system appears to be no more efficient than it was in the 1980s and arguably less so.

Estimates of the cost of bringing a drug to market today range from USD 350 million to USD 1 billion or more. The typical time span for the development of a drug, from target identification to clinical use, is twelve years. As the industry has taken on more difficult disease targets, attrition rates have increased: only one of ten candidate drugs is eventually brought to market. Even this one compound often fails to be efficacious and may, in fact, have substantial adverse effects.

There is a growing consensus that new tools and approaches are needed to facilitate more informed choices throughout the drug discovery and development process. This chapter explores how pharmacogenetics can be used to improve the biomedical innovation process, particularly the translation of early-stage projects into medicines that meet public health needs.

Trends in pharmaceutical innovation

Two indicators of trends in the innovative performance of the bio-pharmaceutical industry are the number of new molecular entities (NMEs) and the therapeutic biological products developed each year. NMEs are potential drugs with novel chemical structures which have never before been approved, in any form, for marketing. Therapeutic biological products are a subset of drugs derived from living material and include – for example – monoclonal antibodies, proteins intended for therapeutic use, immuno-modulators and growth factors. Approval to market a biological product is granted by issuance of a biologics licence. A count of the number of NMEs and biologics licence applications (BLAs) submitted to and/or approved by regulatory agencies provides a preliminary benchmark of likely new drug approvals in the coming years. Across the OECD area, NME and BLA submissions have been on the decline since before the turn of the century.

Figure 3.1 shows the number of submissions and approvals for NMEs at the US Food and Drug Administration (US FDA) over a ten-year period. The decline in approvals of NMEs started around 2000. Although the 2004 figure of 31 NME submissions represents a five-year high (matching the numbers in 2000), it remains far below the record levels seen in the mid-1990s. BLA approvals experienced a decline starting in the late 1990s. Similar trends can be observed at regulatory agencies worldwide.

Figure 3.1. New molecular entities and biological licence applications, US FDA, 1995-2004

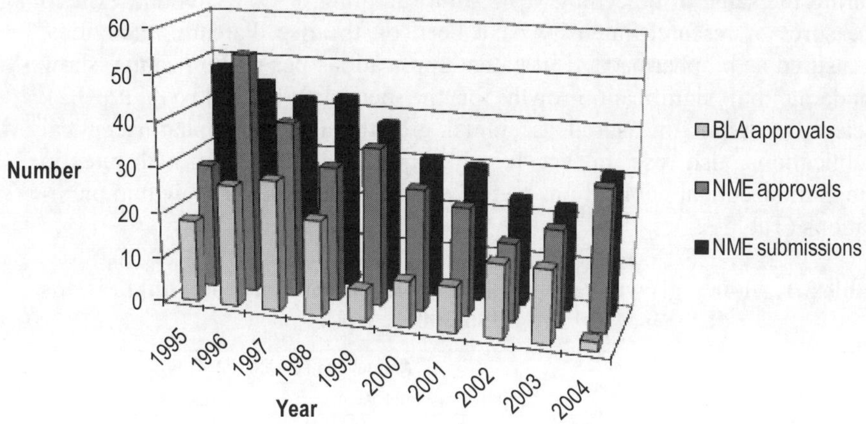

Source: S. Frantz (2004), "Another Long Leaderless Period in Store for FDA", *Nature Reviews Drug Discovery*.

Figure 3.2. Biopharmaceutical patent applications per million capita, 1994-95 and 1999-2000

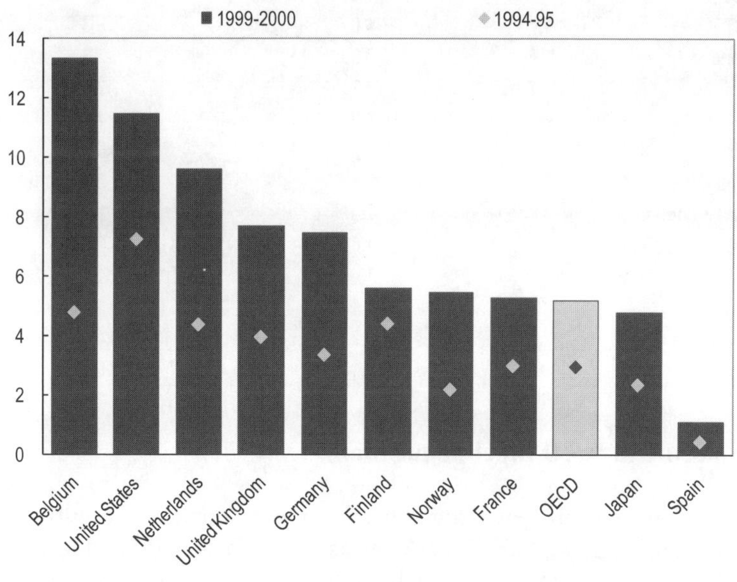

Source: Science Citation Index (SCI).

The slowdown in new drug submissions and approvals cannot be explained by a proportionate decrease in the intensity of biopharmaceutical research during the same or preceding years. In a sampling of OECD countries, two measures of research intensity have been on the rise. Patenting activities, measured as biopharmaceutical patent applications per million capita, show moderate but significant growth for the period 1994-2000 (Figure 3.2). Scientific output, measured as annual growth rates of biopharmaceutical publications, also rose. In fact, for all countries compared, growth rates in biopharmaceutical publications were well above the general growth in publications (Table 3.1).

Table 3.1. Annual growth rates of biopharmaceuticals and total publications, 1996-2006

	Annual growth rate (%)	
	Biopharmaceutical publications	All publications
China	37.6	25.2
Italy	9.5	4.7
Korea	6.6	12.3
United States	5.6	3.5
Belgium	5.6	0.7
OECD	**4.4**	**2.7**
EU27	**4.0**	**3.3**
Finland	3.9	3.6
EU15	**3.7**	**3.0**
Norway	3.3	4.9
Spain	3.0	5.8
Canada	2.5	5.1
United Kingdom	2.0	2.4
France	1.6	0.7
Germany	1.4	1.7
Japan	1.1	1.0
Netherlands	0.7	4.0
World	**6.1**	**5.2**

The pharmaceutical R&D pipeline problem

The slowdown in new drugs entering the market can be attributed to many factors. In 2004, the US FDA released a report regarding the so-called "pipeline" problem, which argues that the applied sciences for development of medical products have not kept pace with the tremendous advances in the basic sciences (US FDA, 2004). The implication is that although scientific advances are accelerating the discovery process, they are not being sufficiently used to guide the development process. More specifically, recent biomedical research breakthroughs have not influenced the ability to identify successful candidate drugs. A further complication is that the path from research to market – even for successful candidates – has become long, costly and inefficient. This is due, in large part, to the current reliance on cumbersome and sometimes poorly predictive methodologies for developing safety and efficacy data.

Given these pipeline problems, scientific creativity and effort must be expended on improving the development process for medical products. An explicit goal should be the creation and implementation of robust development pathways that are efficient and predictable, and that result in products that are safe, effective and available to patients.

A number of initiatives are under way to address pipeline problems. For example, in Europe, the European Commission, together with the European Federation of Pharmaceutical Industries and Associations (EFPIA), developed a proposal for an Innovative Medicines Initiative (IMI) which aims, among other things, to develop streamlined and highly predictive methodologies for developing safety and efficacy data. The US FDA has also launched a number of initiatives, including the Critical Path Initiative.

The US FDA model of the critical path for medical product development is commonly represented as a linear process (see Figure 3.3). At the far left, ideas coming out of basic scientific research enter the prototype design or discovery phase, which is essentially an evaluation process. In drug development, the discovery process seeks to select or create a molecule that exhibits specific desired biological activities. The critical path, which begins when the candidate product is selected for development, comprises three dimensions: assessment of safety, proof of efficacy and industrialisation.

Figure 3.3. The critical path for medical product development

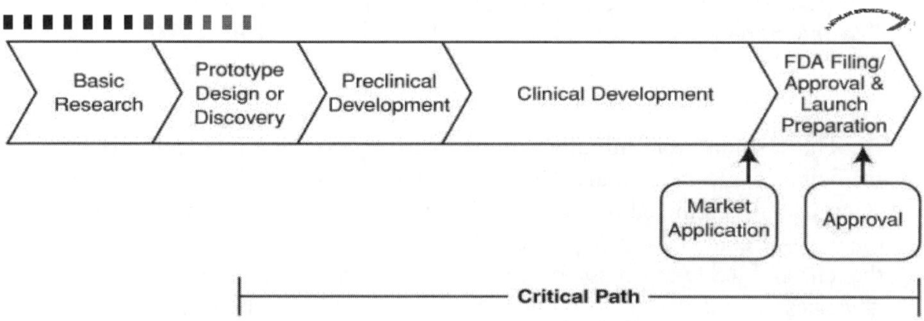

Source: US Federal Drug Administration (2004), "Innovation or Stagnation: Challenge and Opportunity on the Critical Path to New Medical Products".

The US FDA recognises the need to address systemic failures in the pharmaceutical innovation process. Efforts are needed to modernise these three dimensions of the critical path, particularly in terms of targeting the roadblocks or bottlenecks that exist at the interface between dimensions (from the earliest phases of development to commercialisation). The US FDA advocates action to integrate new knowledge and science systems in the R&D process – *e.g.* new animal or computer-based predictive models, biomarkers for safety and effectiveness, and clinical evaluation techniques to improve predictability and efficiency. It also emphasises the need for collaborative efforts by academia, industry and government. These would involve a range of activities, from establishing molecular databases to developing surrogate markers (as discussed in Chapter 2).

The main limitation of the model is the lack of explicit recognition of the interactive and dispersed nature of the medical innovation process. In particular, the model fails to capture the interdependence between the medical industry economy and the medical service economy. It does not take account of the close connections between innovation and the provision, uptake and use of new treatments, a phase at which feedback from purchasers, providers and patients is essential in shaping the innovation process.

The emerging view thus is that improved science is not enough. A systemic approach is needed to address the root causes of current inefficiencies in the pharmaceutical innovation system. To resolve these inefficiencies, such approaches should also take into account downstream factors at the level of the health-care system.

Across the OECD area, a number of large-scale studies and initiatives have recently been published or announced which take this holistic view of the health innovation cycle. These initiatives identify bottlenecks and propose strategies for their removal. Two initiatives of note include the recent European Innovative Medicines Initiative (IMI), *Strategic Research Agenda: Creating Biomedical R&D Leadership for Europe to Benefit Patients and Society* (2005), and US National Institutes of Health (NIH), *Roadmap for Medical Research*.

These studies and initiatives underline the sense that innovation in biomedicine is at a crossroads. On the one hand, advances in the biomedical sciences are predicted to have an enormous impact on the prevention, diagnosis, treatment and cure of disease and disability. On the other, because the current R&D process is costly, unpredictable and inefficient, these advances cannot optimally deliver improved health products and processes to patients. The explosion of promising new fields of research and the corresponding technological opportunities will not automatically translate into broad-based improvements in health care. In order to deliver health improvements, new fields of research and new technologies must be accompanied by changes to the organisation, infrastructure and regulatory frameworks throughout the health innovation system.

These considerations are prompting a re-evaluation of the drug discovery process and a search for solutions to make the innovation process more efficient. A number of OECD initiatives have contributed to addressing this issue. At the OECD, a workshop on health innovation (Berlin, November 2004) referred to the need for a "re-invention of the clinical research and innovation enterprise"; a workshop on international perspectives in pharmacogenetics (Rome, October 2005) considered the impact of pharmacogenetics on biomedical innovation and health care; and a workshop on emerging research models for the delivery of health innovation (Paris, November 2006) discussed alternative models for the development and delivery of health innovation across the OECD area.

Figure 3.4. Pharmaceutical R&D expenditures, 1992-2003

USD millions

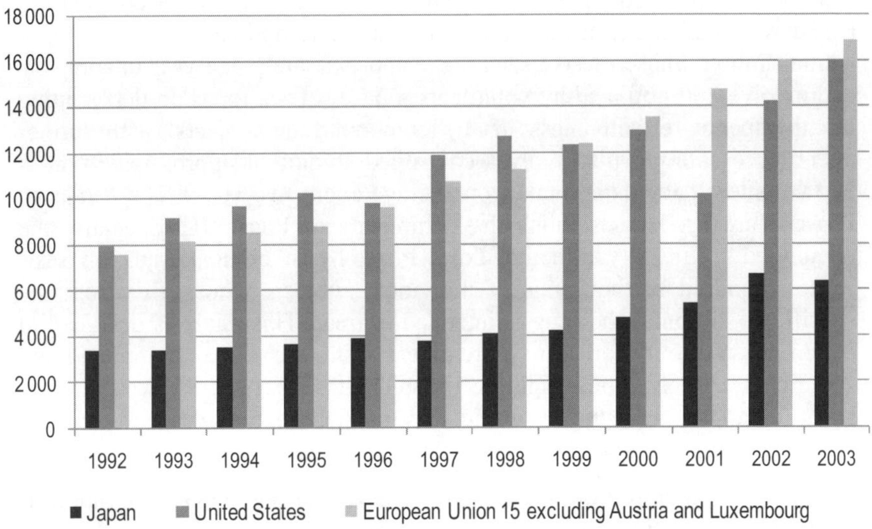

Source: OECD Analytical Business Enterprise Research and Development (ANBERD) database, *www.oecd.org/sti/anberd.*

Pharmaceutical industry R&D expenditure: the high cost of clinical trials

According to the OECD, in 2002 the pharmaceutical industry's R&D expenditure in the United States, the European Union and Japan was significant: USD 16 billion in the United States, USD 17 billion in the EU, and about USD 6 billion in Japan.[1] This capped more than ten years of R&D growth in these regions (Figure 3.4). In large part, R&D expenditure appears to be increasing at a faster rate than the number of innovative drugs generated by the expenditure.

Clinical trials represent one of the largest costs in the drug development process and account for one of the most significant biomedical research expenditures (PriceWaterhouseCoopers, 2005). Before an application is submitted for approval to market a new compound, the developer must demonstrate its safety, quality and efficacy through these trials, which are conducted in three different phases.

Phase I tests study a drug's safety profile, including the potential safe dose range. The studies also determine pharmacokinetics (how a drug is absorbed, distributed, metabolised and excreted) and pharmacodynamics (the duration of its action), as well as possible and optimal methods of drug administration. Phase I tests usually take approximately one year to complete and involve 200-400 healthy volunteers. Phase II tests focus on determining the therapeutic effectiveness of the compound in subjects, with further attention to safety. This involves controlled studies of approximately 200-300 volunteer patients and takes on average about two years. The results of Phase II are used to establish the parameters of Phase III, especially the compound's clinically effective dose. Phase III of clinical trials involves several hundred to thousands of individual subjects who suffer from the specific condition(s) the drug is intended to treat. This phase is designed to determine if the benefits of a treatment with the tested compound are significant enough to outweigh the risks. The tests used in Phase III are the basis for approval of the drug and they must be extremely thorough and meet rigorous standards of safety, quality and efficacy.

The costs incurred during clinical trials reflect the size and duration of the trials. By the time Phase III trials are completed, the costs can be very large indeed. In 2007, the US pharmaceutical industry spent about USD 34 billion in clinical trials; close to 40% of the expenses were incurred for Phase III trials (see Table 3.2).

Table 3.2. R&D by function, PhRMA member companies, 2007

Function	USD millions	Share (%)
Prehuman/preclinical	13 087.4	27.3
Phase I	3 547.7	7.4
Phase II	6 251.0	13.0
Phase III	13 664.7	28.5
Approval	2 413.8	5.0
Phase IV	6 439.9	13.4
Uncategorised	2 498.6	5.2
Total R&D	47 903.1	100

Source: Pharmaceutical Research and Manufacturers of America (2009), PhRMA Pharmaceutical Industry Profile.

The time involved and costs incurred in conducting clinical trials demonstrate that the main problem with pharmaceutical productivity lies downstream from basic research. Recent analysis of drugs under development, provided by the Turku School of Economic and Business Administration, confirms that the share of drugs in clinical Phase III – and thus of the drugs expected to reach the market – is very low, representing only 5-10% of all compounds in the R&D pipeline (see Figure 3.5).

The data also confirm the dramatic reduction in compounds that pass from preclinical to clinical Phase I. This highlights the significant inefficiencies at the interface between these two steps. Many compounds are discarded throughout the drug development process; these are compounds that initially show promise as potential therapeutic agents but are eventually found to be unsuitable. Examples abound of drugs that demonstrate efficacy in early development but do not hold up in more extensive clinical trials. Typically, they either fail to show the predicted benefit in the wider Phase III population or have significant side effects (US FDA, 2004). The further a molecule is taken towards the final marketing authorisation application before its withdrawal, the greater the financial outlay. From the perspective of the pharmaceutical industry, significant savings could be realised by identifying, as early as possible in the development process, those medicines that are likely to fail.

Figure 3.5. Number of compounds in development, by region

As a percentage of total drugs in the pipeline

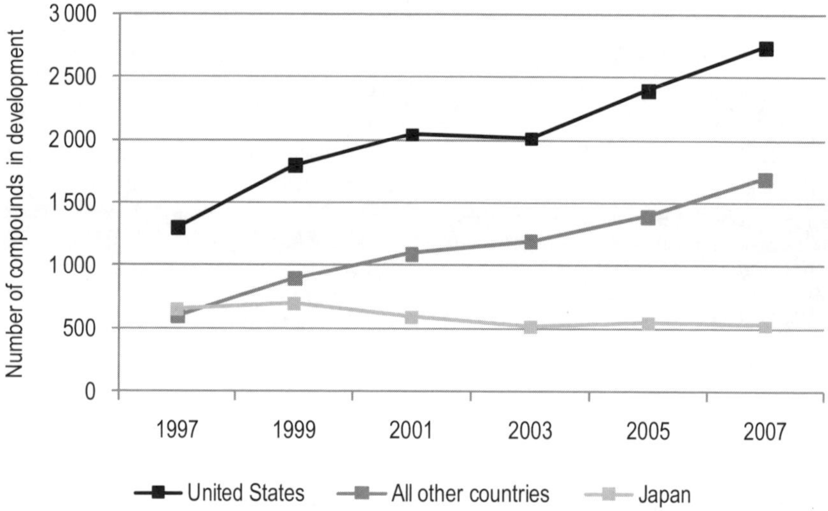

Note: Reflects the number of compounds in clinical trials or awaiting approval as of June of each year. Compounds in development for multiple regions are counted in each region for which regulatory approval is sought, and multiple indications are counted only once.

Source: Pharmaceutical Research and Manufacturers of America, Pharmaceutical Industry, Profile 2009.

Companies have good incentives to reduce the number of medicines that fail in late clinical trials, or even in the post-marketing period. A pharmacogenetics approach could yield significant cost savings over traditional clinical trials.

Reducing the size of clinical trials

As noted above, pharmacogenetics shows potential for improving the efficiency of the drug development process. To recall, genetic biomarkers can be used for two key purposes: to demonstrate efficacy and explain adverse events at a molecular and genetic level; and to identify patient groups that are most likely to benefit from a particular medicine, *i.e.* to "personalise" or "target" the medicine. Using genetic screening to identify potential responder populations prior to clinical trial enrolment could enable demonstration of drug efficacy in a smaller set of subjects. In the case of safety concerns during Phases I and II of drug development, a rapid method of supporting identification of relevant pharmacogenetic biomarkers could create confidence in a go/no-go decision for a commitment to full drug development.

Currently, drug trials are conducted with the assumption that populations are homogenous. In contrast, a pharmacogenetic approach starts by assuming a significant degree of heterogeneity among individuals. This should fundamentally change the way clinical trials are designed and statistically analysed. Indeed, industry information now available suggests that the use of pharmacogenetics could alter the aim of each phase of clinical testing. Phase I could be used to establish proof of concept; Phase II to segment responders, non-responders and adverse responders. The main goal of Phase III would be to refine the results by testing the drug on responders. For example, Phase III could consist of numerous tests on much smaller patient groups, chosen because they exhibit genotypes that suggest they will respond favourably. This progression of data from *post hoc* analysis of Phase I data, through pre-specified clinical trial protocols in Phase II, through prospectively reconfirmed studies in Phase III defines the pharmacogenetics approach to predicting efficacy during drug development. This approach could, in principle, deliver benefits both by reducing the costs of clinical trials for the prospective medicine and by increasing the probability of its success (Roses, 2004).

Using a pharmacogenetics approach, Phase I could involve as few as 50 volunteers. In one particular case, a Phase I study involving only 40 test and 41 placebo patients was sufficient to identify three genetic variants to be used in future patient stratification (Roses, 2004). This approach of selecting smaller groups of participants who are genetically homogeneous for a specific trait is known as "enrichment" (see Box 3.1).

However, the use of pharmacogenetics in trials design is not without its downsides. It should be noted that using enrichment in Phase I clinical trials carries a risk of including only patients who are less genotypically diverse. Thus, while enrichment can be advantageous – particularly in terms of leading to more robust and reliable scientific findings about the patient group for whom the medicine might eventually be prescribed – there may also be a greater risk that rare or delayed side effects might go undetected. Reducing the sample size may carry additional dangers in relation to obtaining important safety information which can be generalised to the population at large or it may lead to a potential overestimation of a drug's efficacy (Moldrup, 2001; Rothstein and Epps, 2001; Thomas, 2001). The benefits and risks of enrichment design studies are illustrated in Box 3.1.

Pharmacogenetic biomarkers may play a more important role in Phase II (*i.e.* during early proof-of-concept studies in a well-defined population). Approximately 70% of new medicines tested in Phase I pass to Phase II. At this stage, trials usually involve 200-300 patients, and are designed to investigate whether the potential efficacy of the new medicine is actually realised. The trials also provide preliminary data on the safety of the new medicine and information about dosage.

Box 3.1. Enrichment: risks and benefits for clinical trials

In conventional, randomised double-blind control trials, no prior account is taken of the response a patient may have to a medicine. This approach carries the risk of masking the efficacy of the medicine in a specific patient population.

When dealing with medicines for which efficacy, dose response or side effects are determined at a genetic level, there is a need to allow for this response in the design of clinical trials. In particular, it is important to ensure that appropriate numbers of patients who are capable of showing a positive response receive the treatment and also that the control group is of an adequate size.

This can be achieved by testing patients for genetics-based biomarkers prior to their inclusion in a trial, and omitting all those who would not respond to the medicine (or are within the at-risk group for adverse drug reactions). Only patients capable of responding to the medicine are then included in the trial, and are randomly divided between placebo and drug groups.

This enrichment approach has the advantage of reducing the size and duration of a clinical trial, thereby bringing an efficacious medicine to market more quickly. However, its disadvantage lies in potentially missing some patients who might benefit, despite their apparent membership in a non-responding group. It may also result in limiting the indications for the medicine. Another issue is that the enriched population may not represent a "real" population in clinical practice, which could make it very difficult to explore a variable dose regime.

Collectively, these factors could lead to the medicine receiving approval for an indication more limited than its true potential, with an associated increase in off-label use (*i.e.* use outside the scope of the drug's approved label). Enrichment is currently being used in clinical trials for a number of medicines. However, there is ongoing debate about the degree to which it is a useful and reasonable approach.

Applying a pharmacogenetics approach may or may not substantially alter the number of patients in Phase II, depending on the complexity of the disease, frequency of the specific genetic trait, and the type of gene action, including the pathway involved.[2] The use of a biomarker as the primary outcome of a pharmacological intervention must be interpreted very cautiously and the number of patients needed for effective Phase II trials will depend on what is necessary to obtain statistically meaningful data on the predictive response profiles.

So far, practical application of the pharmacogenetics approach in Phase II has produced mixed results. For example, using the HER2/*neu* test as a biomarker for selecting patients for cancer treatment reduced both the duration and cost of clinical trials for the drug Herceptin®.[3] However, in trials for the anti-restenosis (the prevention of re-narrowing of a coronary artery after being treated with angioplasty or stenting) drug Tranilast®, investigators were able to identify an SNP variation pattern predictive of adverse events with as few as ten cases, but only when using a control set of 3 000 patients (Roses, 2004). Thus, neither costs nor duration were reduced.

In fact, this mixed picture of early success in some trials and apparently disappointing results in others generally lines up with the two main intended applications of pharmacogenetics. So, employing pharmacogenetics in the study of genetic variants for clinically well-defined diseases (the cancer treatment example) seems to be providing successes, whereas using pharmacogenetics in the study of adverse events (the restenosis example) represents a separate problem which is proving much less tractable. In sum, the use of pharmacogenetics to exclude patients at increased risk represents one of the greatest opportunities, yet one of the greatest challenges.

Generally, rare adverse events are not normally observed before large numbers of subjects have been exposed to a drug. This means that very few patients who experience severe adverse events will be available for studies to further characterise the adverse response and search for associated biomarkers. The search for biomarkers for these patients will necessarily involve fewer patients. This issue involves, therefore, a comparison of a few patients identified by the occurrence of a specific unwanted response to many patients who have not suffered the response. This is a different issue from studying gene variants for a well-defined disease which, as described above for the drug Herceptin, involves hundreds to thousands of equally well-defined patients.

In principle, Phase III presents the greatest opportunities for patient stratification/enrichment through the application of pharmacogenetics. Significant differences in trial outcomes might be expected as well as recommendations that can be translated into clinical practice and risk management. There are, however, some special challenges for pharmacogenetics in such trials. These challenges are primarily logistical. Late-stage clinical trials normally involve subjects from many countries at hundreds of centres, all of which may have their own rules and procedures for the collection, transport, storage and testing of genetic samples. Further harmonisation and/or co-ordination of rules for collecting and handling samples[4] may be necessary, as well as better consensus on reliable statistical approaches for stratification and enrichment of trial populations based on pharmacogenetic data.

Conclusions

The path from research to market for new drugs has become long, costly and inefficient. A number of factors have been evoked to explain this situation. Without doubt, industry is taking on more difficult disease targets, and attrition rates as well as costs have increased.

It also appears that scientific advances are not improving the development process in the way in which they have accelerated the drug discovery process. Many compounds that demonstrate efficacy during preclinical development do not hold up in clinical trials. Significant inefficiencies exist at the interface between target identification and clinical trials. They are often attributed to the methodologies currently employed to deliver the data necessary to assess safety and efficacy. Recent breakthroughs in biomedical research have so far done little to influence the ability to identify successful candidate drugs.

There is thus a growing consensus that new tools and approaches are needed to facilitate more informed choices throughout the drug discovery and development process. These include more robust development pathways that are efficient and predictable and lead to products that are safe, effective and available to patients. Significant savings could be realised by identifying, as early as possible in the development process, medicines that are likely to fail. A re-evaluation of the drug discovery process that is under way may lead to solutions for making the whole innovation process more efficient.

Pharmacogenetics does show potential for improving the efficiency of the drug development process. Specifically, genetic biomarkers can be used for two key purposes: *i)* to demonstrate efficacy and explain adverse events at a molecular and genetic level; and *ii)* to identify patient groups that are most likely to benefit from a particular medicine, *i.e.* to "target" therapy.

However, with the deployment of pharmacogenetics, certain socio-economic and statistical risks will need to be managed. Societies will need to decide how to mitigate risks of increasing genetic exceptionalism or inequity in access once pharmacogenetics enables stratification of patients into responders and non-responders. Using genetic screening to identify potential responder populations prior to clinical trial enrolment could enable demonstration of drug efficacy in a smaller set of subjects. This approach could, in principle, reduce the costs of clinical trials for the prospective medicine while increasing the probability of the success of individual trials. It also drastically changes the traditional view of drug development as a sequential occurrence of well-defined phases. With pharmacogenetics, drug development is being transformed into a data-driven process in which the four clinical phases largely overlap.

Furthermore, the potential of pharmacogenetics to substantially alter clinical trials will depend on the complexity of the disease, the frequency of the specific genetic trait, and the type of gene action, including the pathway involved. Other difficulties may need to be considered and could stand in the way of broader uptake:

- Costs or duration of trials will depend on how many patients are required to obtain statistically meaningful data on the predictive response profiles.
- Biomarker validation studies are difficult and costly. These are large-scale prospective studies that measure genetics and other biomarkers over time and follow up well-defined cohorts of patients.
- Enrichment can be advantageous, particularly in terms of leading to more robust and reliable scientific findings about the patient group for which the medicine might eventually be prescribed. However, there may also be a greater risk that rare or delayed side effects could go undetected.
- Late-stage clinical trials normally involve subjects from many countries and hundreds of centres, all of which may have their own rules and procedures for the collection, transport, storage and testing of genetic samples. This implies the need for better harmonisation of rules in these areas.

Despite these difficulties, evidence demonstrates that pharmacogenetic biomarkers have the potential to be useful in early clinical trials (perhaps especially in Phase II, where figures suggest that pharmacogenetics is used in 20-30% of cases, with a higher proportion in oncology) as well as – so long as the logistics are sorted out – in larger Phase III trials in which enrichment and stratification may make the difference between overall success or failure. Perhaps most importantly for the way in which innovation in medicine is likely to develop, pharmacogenetics is making a reality of the conceptual "virtuous cycle" of innovation (see Figure 1.1) and promises to bring to an end the inefficient and unrealistic linear model of innovation that pervaded the late 20th century. But to succeed in this new model, new more "open" approaches to sourcing data and materials will be essential.

Notes

1. The OECD R&D reporting focuses on who performs the R&D activity. It therefore covers only R&D expenditures in the internal sector for activities executed by the pharmaceutical industry itself (intramural expenditure).

2. Complex processes beyond drug metabolism may contribute to individual susceptibility to a drug. This means that a single genetic trait (or biomarker) associated with an adverse drug reaction (ADR) or other clinical end point may constitute a risk factor, but it may be neither necessary nor sufficient to produce the ADR by itself. There may be multiple mechanisms of toxicity, such as multiple pathways of drug elimination, contribution to toxicity by metabolites as well as the parent drug, and potential gene-gene interactions.

3. See Press and Seeling (2004).

4. For details, see OECD (2007).

References

European Innovative Medicines Initiative (IMI) (2005), "Strategic Research Agenda: Creating Biomedical R&D Leadership for Europe to Benefit Patients and Society", *http://ec.europa.eu/research/fp6/pdf/innovative_medicines_sra_final_draft_en.pdf.*

Frantz, S. (2004), "Another Long Leaderless Period in Store for FDA", *Nature Reviews Drug Discovery*, p. 289.

Moldrup, C. (2001), "Ethical, Social and Legal Implications of Pharmacogenomics: A Critical Review", *Community Genetics*, 4, pp. 204-214.

OECD (2003), Analytical Business Enterprise Research and Development(ANBERD) database, *www.oecd.org/sti/anberd.*

OECD (2007), *OECD Guidelines for Quality Assurance in Molecular Genetic Testing*, OECD, Paris.

PhRMA (Pharmaceutical Research and Manufacturers of America) (2009), Pharmaceutical Industry Profile, p. 53.

Press, M.F., and S. Seelig (2004), "Lessons Learned From the Development of a Diagnostic to Predict Response to Herceptin", Targeted Medicine Conference Report.

PriceWaterhouseCoopers (2005), *Personalised Medicine: The Emerging Pharmacogenomics Revolution.*

Roses, A. (2004), "Pharmacogenetics and Drug Development: The Path to Safer and More Effective Drugs", *Nature Reviews Genetics*, 5, pp. 645-656.

Rothstein M.A. and P.G. Epps (2001), "Ethical and Legal Implications of Pharmacogenetics", *Nature Reviews Genetics*, 2, pp. 228-231.

Thomas, S.M. (2001), "Pharmacogenetics, The Ethical Context", *The Pharmacogenomics Journal*, 1, pp. 239-242.

US FDA (US Food and Drug Administration) (2004), "Innovation or Stagnation: Challenge and Opportunity on the Critical Path to New Medical Products", *www.fda.gov/oc/initiatives/criticalpath/whitepaper.html.*

Chapter 4

Business Models for Pharmacogenetics

This chapter discusses the challenges posed by pharmacogenetics to established "big pharma" business models in the pharmaceutical and diagnostics and device industries. It considers the viability of current business models and the financial incentives of different types of firms.

The pharmaceutical industry has traditionally adopted a "one drug fits all" approach to research and development. It targets therapies to the broadest patient population that might possibly benefit and relies on statistical analysis of this population's response to predict therapeutic outcomes in individual patients. Pharmacogenetics is challenging this "blockbuster" business model. In excluding, as much as possible, variability of responses to drugs due to genetics, pharmacogenetics is likely to result in smaller markets for many individual medicines, even such widely used drugs as Warfarin® (Washington Post, 2007).

Device and diagnostic firms are also under pressure. By definition, pharmacogenetics requires the co-development of a drug and a test or assay. Device and diagnostic makers have had difficulty taking the lead in developing pharmacogenetic assays because genetic tests traditionally have had low returns and because of the difficulty in establishing intellectual property rights for biomarkers. Co-development could result in increasing competition from pharmaceutical firms for the development and marketing of pharmacogenetic tests.

This chapter reviews the financial incentives that could shape the path of pharmacogenetic products – both drugs and associated genetic diagnostics – to the market. It explains some of the roadblocks ahead in two key industries, and the levers policy makers can use to improve the innovation environment.

The blockbuster model and pharmacogenetics

In the pharmaceutical sector today, the strongest indicator of company profitability is market leadership in individual therapeutic categories of drugs. Most pharmaceutical companies rely on "blockbuster" products to establish and maintain market leadership. Blockbusters are medicines with peak annual sales in excess of USD 1 billion; they usually address a previously unmet health need and are targeted to the general population or large subsets of it. Thus, the pharmaceutical industry is built on a business model in which a small number of blockbuster products generate the majority of company revenues and therefore dictate the overall direction and clinical focus of company strategies.

Reliance on the blockbuster model has led to a reduction in the number of medicines that make it to market. Companies will withdraw from developing medicines that appear to have limited application in favour of those with blockbuster potential. The model encourages the development of life-style medicines, which can be marketed to a general public, and discourages the development of treatments that address small populations.

The blockbuster model also encourages the development of "me-too" compounds; that is, several companies develop similar therapeutics that are targeted at the same aspect of a disease and generally rely on the same mechanism of action (*e.g.* statins, ACE inhibitors, H2-antagonists, or proton pump inhibitors). The advantage of me-too drugs is that they can be developed rapidly and usually result in a compound with improved clinical benefits over the original drug. This reduces the period during which the first-to-market company enjoys market exclusivity. Competition from generics also has an impact on company strategies. These factors ultimately affect the pricing and marketing of pharmaceutical products and therefore contribute to a business model that favours compounds that command high prices across large markets.

The blockbuster model for drug discovery and development poses a potential limitation to the uptake of pharmacogenetics. Most pharmaceutical companies may be reluctant to embrace an R&D approach that could imply both a fragmentation of current large markets and smaller markets for each individual product. Under the blockbuster model, a potentially limited market is one of the reasons for discarding a potential therapeutic, particularly in late-stage clinical trials. It should be noted, however, that some firms – such as Genzyme – have adopted a business model that targets the disease and patient sectors ignored by the blockbuster model (see Box 4.1).

Box 4.1. Incremental cost-effectiveness ratio and the value of pharmacogenetics: a hypothetical example

In economics, cost-effectiveness refers to the comparison of the relative expenditure (costs) and outcomes (effects) associated with various (two or more) courses of action. It is typically expressed as an incremental cost-effectiveness ratio (ICER), which measures the change in costs over the change in effects (a low ratio implies that increased effectiveness is achieved at relatively low cost, and so is desirable):

$$\text{ICER} = \frac{\text{change in costs}}{\text{change in effect}}$$

Here, ICER measures the effect of switching interventions – *i.e.* to pharmacogenetics and the co-development of medicines and devices – rather than the standard economic use of "incremental" which is the effect of an additional unit of a specific measurement (*e.g.* the effect of an additional dollar spent on a public health campaign).

Many health-care systems employ an ICER threshold. That is, they identify a dollar figure with the ratio and approve only those medicines or therapeutics that can deliver good value at or below the identified price.

…/…

Box 4.1. Incremental cost-effectiveness ratio and the value of pharmacogenetics: a hypothetical example *(continued)*

In this hypothetical scenario, which uses rounded figures for simplicity, a drug developer identifies a population of 100 people with a certain disease. It begins to develop a new therapeutic, but does not yet know what the response rate will be. Early estimates demonstrate that the incremental cost of introducing the new drug to the health system will be approximately USD 10 000 per patient. The expected health gain for patients who respond is 0.5 life-years.

The drug developer estimates that its revenues will be USD 10 000 per patient, and that the cost of the genetic test is USD 100. Figure 4.1 shows the incremental cost-effectiveness ratio (ICER) for two alternative scenarios. In Scenario 2, a test is used to identify respondents and thereby to target the new therapeutic (the assumption is that 10% of patients are targeted following tests). In Scenario 1, no test is administered; the therapeutic is given to all 100 patients.

The horizontal axis of Figure 4.1 indicates response rate, from very low to very high (100%). In Scenario 1, as the response rate increases, the incremental cost per life-year gained (the vertical axis) is reduced. This is because more people respond to the treatment, which results in more health gains at equal cost. Hence, without a test, the incremental cost per life-year gained is very high with a low response rate, but obviously becomes lower as response rates increase. In Scenario 2, in which a test is available to predict patient response, the ICER is stable: the only costs incurred are for the treatment of people who have been shown to be genetically pre-disposed to benefit from that treatment.

Revenue for the pharmaceutical company differs significantly between the two scenarios. Under the no-test scenario, the company receives revenue for each of 100 people treated. Under the with-test scenario, the company stands to gain revenue from only a fraction of the total population with the disease in question. When response rates are low, there revenue to be gained is limited. However, there is convergence when response rates move towards one.

From the pharmaceutical industry perspective, this simplified example demonstrates the lack of incentive to develop pharmacogenetic products. From the public perspective, however, it shows the potential gains to be realised in developing and introducing this novel approach. Tests to predict response rates could lower the ICER and reduce overall health system costs and improve efficiency as well as patient satisfaction.

This highlights the need to examine the R&D incentives in jurisdictions that employ a minimum ICER threshold to determine reimbursement of pharmaceuticals. For example, if the jurisdiction employs an ICER threshold of USD 50 000 per life-year gained, the company knows that it will not receive revenue for a new therapeutic unless it can achieve an ICER below this rate. In the hypothetical example, this is threshold is realised at a response rate of 0.4. Under the no-test scenario, company revenue would be zero for response rates lower than 0.4, simply because they will not be reimbursed at ICER higher than USD 50 000. By contrast, if the company develops a test to identify responders, it can reduce the ICER and generate revenue at response rates lower than 0.4. This suggests that in jurisdictions with ICER thresholds, there may be an incentive for companies to develop genetic tests for therapeutics with low response rates.

Figure 4.1. Hypothetical example of incremental cost-effectiveness ratio, company revenues and therapeutic response rates

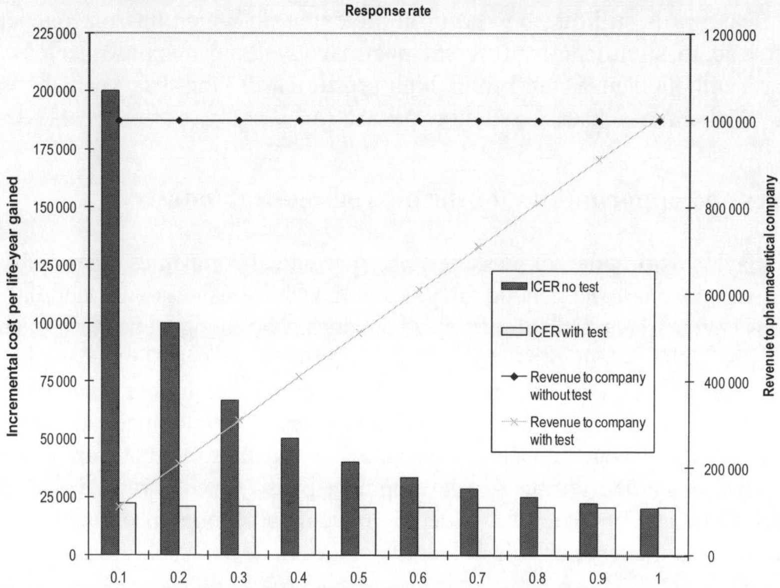

What is unclear, at present, is whether pharmacogenetics will actually push the pharmaceutical sector towards adopting a new business model for drug development. The push for a new approach might depend to some extent on whether individual responses to drugs vary because of differences in pharmacokinetics or pharmacodynamics.

- Pharmacokinetics describes the course of drug and metabolite levels in different tissues and can be viewed as the speed with which an individual metabolises a drug. Differences in pharmacokinetics between patients imply differences in drug dosage.

- Pharmacodynamics describes where (the site/target) and how (the mechanism) a drug acts on the body. These responses are in part determined by genetic variations among individuals. As a result, a patient may show no response to a drug, exhibit the predicted clinical response or experience adverse effects. A difference in pharmacodynamics among patients implies different treatments.

If pharmacogenetics reveals, for the most part, differences in pharmacokinetics, getting the dosage right for individual patients may not require a drastic change in the present business model. Pharmacogenetics will lead to better health outcomes through more accurate dosage regimes. However, if pharmacodynamics segments patient cohorts into responders and non-responders, the blockbuster model is under much greater pressure because different drugs may be necessary for a same disease.

The jury is still out, however, on the extent to which pharmacogenetics will lead to significantly different pharmacodynamic responses across the board, although in a number of high-profile cases, this has been the case (*e.g.* Alzheimer's disease, oncology, Warfarin®).

Challenges and opportunities for the pharmaceutical industry

Laying aside external pressures, pharmaceutical companies are only likely to invest in pharmacogenetics if there is a viable business case and/or if it opens new markets. One of the key advantages pharmacogenetics brings to the industry is the potential to reduce drug development costs. For example, the early identification of biomarkers could ensure that a higher proportion of drug leads actually make it through all the phases of development and onto the market.[1] Pharmacogenetics might also allow companies to pursue drugs which they had discarded during the development phase or withdrawn because of lack of efficacy or toxicity for specific patient sub-groups if enrichment and stratification make it possible to identify and exclude such sub-groups.

Therefore, pharmaceutical companies might significantly improve their drug development capabilities by using pharmacogenetics to increase the understanding of the molecular basis of disease and drug response and they might also improve the efficiency and effectiveness of clinical trials.

Pharmaceutical companies may also see an opportunity to diversify into new areas. With pharmacogenetics, the service component of health care becomes increasingly important. Pharmaceutical companies would no longer simply push high volumes of medicine through standard distribution channels. Rather, drug discovery and development become more closely linked to managing disease, customising treatments, and offering genetics-based diagnostics and treatments. The pharmaceutical industry might respond to this service component of pharmacogenetics by, for example, developing new distribution channels and operational capacities in order better to meet the needs of physicians and patients.

It should be recognised, however, that not all therapeutic areas will be equally suited to a pharmacogenetics approach. In some cases, clinically useful biomarkers may not exist, or new strategies drawing on pharmacogenetics biomarkers may not offer significant benefits over traditional therapy. Pharmaceutical companies may therefore apply pharmacogenetics on the few particularly promising therapeutic areas most amenable to a pharmacogenetics approach. Currently these are oncology, cardiovascular diseases, infectious diseases, the central nervous system and transplants (these disease areas include some of the top priority areas for many OECD member countries).

Among these therapeutic areas, oncology is the area in which pharmacogenetic approaches are currently used most. Cancer is a disease of DNA, with many known variations based on different genetic modifications in somatic cells. Rapid and accurate cancer diagnoses are needed to establish appropriate treatment. Oncologists are interested in using genetics-based biomarkers in diagnosis and prescription. Pharmaceutical companies therefore have a strong incentive to develop medicines targeted at appropriate patient groups. Oncology is likely to provide an important model, or at least demonstration, of how pharmacogenetics can influence the innovation process for new medicines. It will also provide an example of how the industry might transform its distribution and operational capacities to become both a product and service provider.

A likely implication of a (probably accurate) perception of the differential utility of pharmacogenetics for different therapeutic targets is that different companies may embrace pharmacogenetics to different extents and at different rates, depending on their areas of therapeutic specialisation. It is too early, however, to predict the extent to which such differential uptake may lead to disruption or even stratification within the existing predominant model or – just as importantly – lead to divergence in the overall innovative efficiency and effectiveness between significant pharmacogenetics adopters and non-adopters.

To summarise, it is unclear whether a radical new business model for drug development is in the offing, even if there is an opportunity to take advantage of population variations in drug response. Currently, incentives for the pharmaceutical industry to invest in pharmacogenetics and targeted therapies are mixed, given the high cost of the basic research and the uncertain commercial success of co-marketing a diagnostic and drug.

Governments, regulators and private health insurers may have an important role to play in making the uptake of pharmacogenetics more attractive. In many countries one policy lever under their control, to be discussed in Chapter 6, is the ability to adjust the pricing and reimbursement of targeted therapies and of their related diagnostics in order to reflect the expected increased health benefits or value per patient. However, even if prices and/or reimbursement practices are adjusted in this way, the patient populations may be too small to make commercial development viable. To make some of the smaller markets more attractive, it may be necessary to introduce measures that draw on experience with orphan drug legislation, which was instituted to attract investment in and development of medicines for rare diseases.

SMEs and the biotechnology sector

Irrespective of the influence on overall business models, the pharmaceutical industry is increasingly interested in integrating the consequences of pharmacogenetics into their R&D process even if they do not do this in house. A recent study (Webster *et al.*, 2004) indicated that between 1997 and 2004, the industry established a significant number of alliances and collaborations in pharmacogenetics and pharmacogenomics. The study identified 47 small and medium-sized enterprises (SMEs) that develop pharmacogenetics. This small group of companies, located primarily in North America (29 firms) and Europe (18 firms), provides services to support drug discovery. Recognising that pharmacogenetics can be applied along the entire R&D pipeline, these companies have very diversified portfolios. The companies developing technologies to support genotyping focus almost entirely on developing diagnostic tests as distinct products or kits, rather than providing in-house testing services. They focus on a variety of products and technological options, including:

- services and products to support preclinical and clinical drug development (*e.g.* genotyping services or testing of the biopharmaceutical characteristics of lead compounds;
- toxicity screening;
- genotyping and association studies and products such as mutation detection kits, assays and chips;
- databases on adverse drug reactions;
- software tools.

High company failure rates are a striking feature of this sector: 18 firms listed in 2002 were no longer active by 2004. Based on the relatively small number of dedicated SMEs and the high turnover, researchers conclude that most companies see pharmacogenetics as an additional tool in the drug development toolbox. So far, mainly diagnostic companies see a "pure" market for pharmacogenetics products.

The device and diagnostics industry

The medical device and diagnostics industry includes a sub-segment of companies which provide genetic or genetics-related tests to support the diagnosis of disease as well as information relating to therapeutic regimes. This industry sector has a different business model from that of the pharmaceutical industry, in particular for the part of the industry focused on test development and marketing. Most device companies do not undertake research to identify the biomarker for which tests are developed. Historically, intellectual property and competition in this sector has centred on platforms (*i.e.* the instrument used to measure analytes). The profit margins on single tests are lower than for pharmaceuticals, with little or no scope for premium prices. Such companies usually rely on volume sales and may be extremely innovative in the way they develop and apply their platforms.

Traditionally, the device and diagnostics industry has supplied assays of clinical value and utility with little competition from the pharmaceutical industry. This worked for two reasons: adverse reactions could be detected with common tests for which there was no available intellectual property; and there were rather low profit margins on such tests and assays.

The emergence of pharmacogenetics is changing this situation. Today, biomarkers that can be used to measure drug efficacy are identified during the drug discovery and development process. These include biomarkers for patient stratification, as well as novel biomarkers for adverse side effects. Increasingly, these biomarkers have a genetic basis: they either identify a particular genotype or indicate the presence of a particular protein (or other entity measured) which is, in turn, the product of a specific DNA sequence. Such biomarkers are used in clinical trials to identify patients who will respond appropriately, adversely or not at all to a medicine. They can also be used to identify appropriate dose ranges and the clinical endpoint, or the desired outcome, of a treatment in a clinical trial. For those medicines that are eventually brought to market for a genotype-specific population, biomarkers may be the necessary first step in defining which patients are treated, while ensuring that the therapeutic benefit is offered to those who will benefit most.

This use of biomarkers represents new opportunities for the device industry, but also a new threat. The drug development companies will need to have these assays in place as they move into clinical trials. It is therefore likely that the drug companies will expand beyond straight pharmaceuticals to develop – and even market – devices and diagnostics, perhaps through partnerships or company acquisitions. But even if pharmaceutical and device companies collaborate to co-develop the medicine and the assay, a number of significant hurdles remain for bringing targeted therapies and accompanying diagnostics to market.

Challenges in co-development of drugs and tests

Pharmacogenetics could stimulate the development of a new business model, one in which pharmaceutical and device industries co-develop drugs and devices. But the process of developing, in parallel, a device or diagnostic and a related drug poses its own challenges. Traditionally, these two processes advance asynchronously. It is not until the biomarkers for a given therapeutic intervention are known and the appropriate association studies are done that there is a real opportunity to develop independent assays that will provide reliable information in a clinical setting. Under any new model, these activities would need to be co-ordinated.

At present, regulation of the combined use of a diagnostic and a therapeutic is evolving slowly and unevenly across OECD countries. There is concern that an already complex approval process will become even more complicated. Also, current pricing and reimbursement systems do not adequately recognise the value of the contribution made by the device industry. Identifying patients who will respond well to pharmacogenetic medicines holds potential financial benefits for the overall health-care system (see Boxes 4.1 and 4.2). However, this financial gain is generally not captured by the diagnostics company. In both the United States and Europe, pricing and reimbursement for diagnostic tests is essentially cost-based; it does not reflect the clinical value the test might offer the patient or the economic value to the health-care system. The lack of recognition of the benefit of testing within the pricing and reimbursement mechanisms represents a genuine disincentive to the device industry to develop new, genetics-based assays.

Some significant policy thinking needs to be done by governments about how to remove perverse disincentives to capture the efficiencies offered by pharmacogenetics. These might include issues of labelling, market access, sharing of intellectual property and who captures what value in a partnership. If further concentration of market power – and the associated inefficiencies – are to be avoided, policy mechanisms will be needed to encourage partnerships across the sectors (big pharmaceuticals, devices and diagnostics, biotechnology and informatics, SMEs, etc.) and create favourable conditions for more open innovation to evolve.

Box 4.2. The economic value of pharmacogenetics

Health-care systems stand to derive greater value from the pharmacogenetics approach. The combination of a new therapeutic medicine and a genetics-based diagnostic test which makes it possible to identify patients who will respond to this medicine holds greater potential than the new therapeutic alone.

The following scenario highlights the use and economics of a new therapeutic agent (Tx) with or without a diagnostic test (Dx):

Tx with no Dx	Tx associated with a highly sensitive and specific Dx
100 patients receive Tx	100 patients are tested
20% respond	20 receive Tx
Willingness to pay: USD 1 000	Willingness to pay: USD 6 000
Total value generated: 100 x USD 1 000 = USD 100 000	Total value generated: 100 x 0.2 x USD 6 000 = USD 120 000

Use of a diagnostic test increases both the precision of the diagnostic decision and the possible price of the therapeutic agent.

However, this increased financial gain raises the question of who should capture the value of bringing together the therapeutic and the diagnostic test. A reimbursement regime is needed which provides appropriate incentives to the different components of the system, allowing for fair financial benefit. Without this, society's perception will be that the pharmacogenetic testing approach is not cost-effective.

Ultimately, the question of who will capture the value of a linked diagnostic-therapeutic depends on many factors:

- Pricing and reimbursement constraints.
- Intellectual property protection.
- Competitive market conditions.
- Timing of entry.
- Insurance market competitiveness.
- The characteristics of the diagnostic and therapeutic products.

Along with scientific and clinical considerations, whether, when and how value will be created is inextricably related to who captures it. At present, the low level of reimbursement for genetic tests is a significant disincentive for the device industry to engage in the co-development of pharmacogenetic assays.

Source: Adapted from Garrison (2005), "Co-development and Marketing of Pharmacogenetic Tests and Therapeutics: Economic Incentives and Policy Implications", presentation at the OECD workshop "An International Perspective on Pharmacogenetics: The Intersections between Innovation, Regulation and Health Delivery", Rome, Italy, 17-19 October, *www.oecd.org/sti/biotechnology/pharmacogenetics*.

Creating and capturing value from pharmacogenetics

Who might capture the benefits associated with the use of pharmacogenetics – and, indeed, whether any benefits are captured at all – remains a crucial question, and one that policy can influence significantly.

Two key parameters are worth drawing particular attention to – the potential response rate of a new therapeutic for which genetics plays a role in affecting response, and the willingness to pay for clinically effective outcomes.

The potential response rate to a new therapeutic will be a key parameter in the decision to invest simultaneously in research and development of a pharmacogenetics-related test. Box 4.2 illustrates this concept with a hypothetical example, which also suggests the need for policy intervention to establish incentives for developing pharmacogenetics. As shown in this simplified example, the incentives are likely to differ from one health-care system to another and depending on the methods used to determine coverage decisions. Many health-care systems currently cover only services and treatments whose cost-effectiveness ratio is equal to or less than a specific cut-off (threshold) value. The example highlights the need to examine the various types of incentive signals that flow between the health-care system and the R&D community.

Innovation in pharmacogenetics can be slowed by a perception of weak financial incentives. The means by which government and private health insurers operate pricing and reimbursement systems will create incentives or disincentives for innovation in this area. Policies need to be developed that recognise the potential value of diagnostic tests combined with targeted therapy to the health-care system as a whole and to patients in particular.

The potential benefits to a targeted group of patients, predictable in advance, may mean that there is a greater willingness to pay for interventions for which there is a very high response rate than for those that may provide significant numbers of non-responders or poor responders. Reimbursement policy could be used better to reward innovations that deliver the relatively higher clinical efficiency associated with a high response rate. Indeed, some have advocated the shift of reimbursement policy in light of the advent of pharmacogenetics to a much more value-based reimbursement system. The example in Box 4.2 illustrates such a "value-based" reimbursement approach, with a greater willingness to pay for interventions with high clinical effectiveness.

The main difficulty with the example in Box 4.2 is that the assumed greater willingness to pay implies that the health system can tolerate an increase in the aggregate costs of reimbursement either by government or other health-care provider/insurer. Increased aggregate costs for an intervention seem unlikely unless associated benefits outweigh such incremental costs and would not otherwise be derived. In real-world situations this may or may not be the case – in any event, health technology assessments and associated decision making will need to take account of any such externalities.[2]

Of course, as discussed in Chapter 3, the effective harnessing of pharmacogenetics promises improvements in the efficiency and effectiveness of the drug discovery and development process. It could, in principle, mean more products entering the marketplace than has recently been the case. These increases in efficiency – and increases in the numbers of effective products entering the marketplace – may do as much as or more than any changes in reimbursement and coverage policies to influence the development of industry business plans and ensure the capture of value.

In principle, pharmacogenetics offers opportunities for capturing new value from many existing medicines for uses which without population stratification seemed relatively ineffective (and perhaps were not reimbursed) and/or for near-miss products on the shelves of pharmaceutical companies which did not enter the market owing to poor efficacy or safety issues in non-stratified trials. However, the application of pharmacogenetics to existing medicines may also remove value, if the market size for existing medicines diminishes as a consequence of new knowledge. Incentives for applying pharmacogenetics to existing medicines by any but the public sector therefore seem weak. If regulatory levers are used to give the private sector an incentive to revisit existing products, some form of risk sharing, such as public-private partnerships, may be necessary.

Finally, new orphan categories of disease may be created with effective therapies for some disease polymorphisms but not for others (*i.e.* where pharmacogenetics splits patients into responder and non-responder groups for specific therapies). Besides the obvious equity and ethical issues associated with such stratification, policy makers are likely to be challenged to create mechanisms through which the associated market failures can be addressed. The options for doing this – for example, though public-private partnerships – need to be developed.

Conclusions

Widespread adoption of the pharmacogenetics approach is largely dependent on its uptake by the pharmaceutical and device industries. The private sector's perception of the financial opportunities this technology offers could hamper progress. Understanding the incentive environment for health innovators is vital to harnessing the benefits of advances in pharmacogenetics research for public health.

Recent trends in alliances demonstrate that the pharmaceutical industry is interested in integrating pharmacogenetics into the R&D process. Yet the pharmaceutical blockbuster business model may limit the development of pharmacogenetics-based drugs. By enabling market stratification (either by drug or by patient group), pharmacogenetics may, in fact, restrict the market size for a drug. This could represent a fundamental change in the dynamics of the pharmaceutical industry. Rather than relying on a single blockbuster drug, pharmacogenetics might result in a more diversified portfolio of drugs which are individually less profitable than a "blockbuster" but may yield greater health improvements.

In the long term, a change in the pharmaceutical industry's business model may be driven by reduced attrition and more successful development due to pharmacogenetic products. More moderate revenues from a wider range of approved pharmacogenetic products (*versus* a single blockbuster) may ultimately be commercially viable. Pharmacogenetics may also identify new pathways or useful targets for the development of new classes of medicines. The capacity to define by genotype patient groups that will respond positively to a given drug may also allow pharmaceutical companies to pursue drugs previously discarded during the development phase or withdrawn post-marketing because of toxicity to specific patient sub-groups or uncertainties regarding efficacy. Ultimately, this approach may improve clinical outcomes and add greater value to the health-care system as a whole.

In re-aligning to support pharmacogenetics, the device industry faces a specific set of challenges and potential benefits. Testing devices are typically developed after a target has been identified; device companies are little involved in target identification and in matters related to revenue or protecting market share through intellectual property rights. Because of low profit margins, companies rely on volume sales and competition comes largely from other companies that develop faster or better devices. In reality, current pricing and reimbursement schemes (which are essentially cost-based) do not adequately reflect the value of the device industry's contribution to the patient or to the health-care system.

There is good reason to believe that pharmacogenetics will require a business model in which drugs and devices will need to be developed in parallel. This creates certain challenges:

- Traditionally, the processes of drug and device development have been asynchronous. The need to co-ordinate these activities raises questions as to whether pharmaceutical companies will partner with device companies or will seek to develop their own in-house expertise.

- At present, regulation of the combined use of a therapeutic and a diagnostic is evolving slowly and unevenly across the OECD countries. There is concern that an already complex approval process will become even more complicated.

- Within current pricing and reimbursement mechanisms, the lack of recognition of the true value of testing discourages the device industry from engaging in developing new, genetics-based assays.

These issues suggest that the pursuit of pharmacogenetics will need to involve the development of new business models. Action on the part of governmental and regulatory bodies as well as private health insurers may well be needed to catalyse this change in areas such as pricing and reimbursement; recognition of the value of diagnostic tests to the health-care system as a whole; and, possibly, mechanisms for capturing and protecting intellectual property from biomarkers.

Notes

1. A major focus of initiatives such as the European Commission's Innovative Medicines Initiative and the FDA's Critical Path Initiative.

2. See Chapter 8 of OECD (2005) for a further discussion of these issues.

References

Garrison, L. (2005), "Co-development and Marketing of Pharmacogenetic Tests and Therapeutics: Economic Incentives and Policy Implications", presentation at the OECD workshop "An International Perspective on Pharmacogenetics: The Intersections between Innovation, Regulation and Health Delivery", Rome, Italy, 17-19 October, *www.oecd.org/sti/biotechnology/pharmacogenetics*.

OECD (2003), Analytical Business Enterprise Research and Development (ANBERD) database, *www.oecd.org/sti/anberd*.

OECD (2005), *Health Technologies and Decision Making*, OECD, Paris.

Washington Post (2007), "For the First Time, FDA Recommends Gene Testing", 17 August.

Webster, A., P. Martin, G. Lewis and A. Smart (2004), "Integrating Pharmacogenetics into Society: In Search of a Model," *Nature Reviews Genetics*, 5, pp. 663-669.

Chapter 5

Physician and Patient Demand for Pharmacogenetics

This chapter reviews current evidence of physicians' attitudes to pharmacogenetics. It also covers patients' perspectives and possible ethical and equity issues, particularly those concerning developing countries. It closes with consideration of the opportunities and challenges for incorporating pharmacogenetics in clinical care as well as potential organisational and system-wide hurdles. This will necessarily create both technical and educational challenges for health professionals. Successful integration of pharmacogenetics will depend upon appropriate infrastructure and trained personnel.

Figure 5.1. Per capita expenditure on pharmaceuticals, 1995-2005

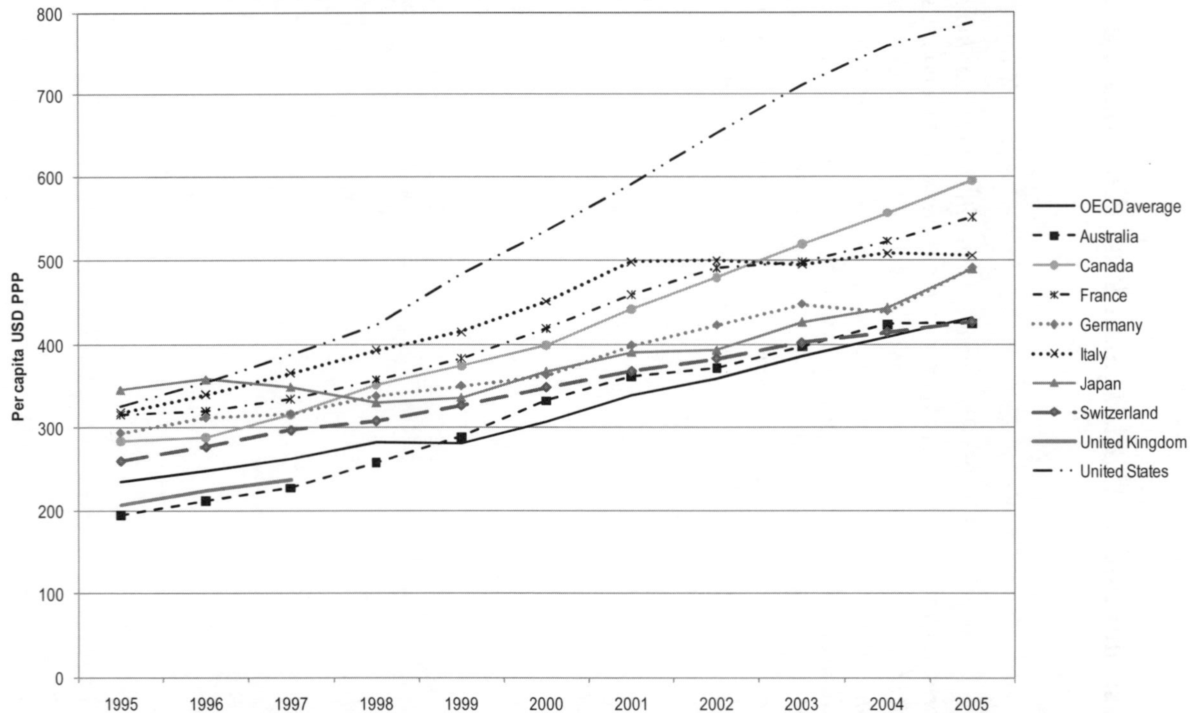

Source: OECD Health Data 2008 (www.oecd.org/document/16/0,3343,en_2649_34631_2085200_1_1_1_1,00.html).

Pharmacogenetics has the potential to increase the efficiency of health care for patients by reducing the risks associated with drug treatments. In particular, patients may face fewer risks of adverse events because the variability of their response to drugs is taken into account.

Pharmacogenetics aims to increase the probability that the best possible therapeutic decisions are made for a patient. In many instances, however, the information gained from pharmacogenetic assays is not easy to interpret. Most diseases are complex and may involve more than one gene with variable penetrance. The behaviour of these genes also depends on environmental factors such as age, sex and diet. Hence, pharmacogenetics generally provides probabilistic rather than absolute information. In prescribing, pharmacogenetic assay results will have to be utilised in conjunction with other medical information.

This chapter discusses the pharmacogenetic end users in the health care system, how they get their information about targeted treatments, and what motivates them to demand these treatments. The chapter considers some of the implications of pharmacogenetics for the organisation of health-care delivery systems and the wider issues that may influence the performance of these systems. It also discusses how the availability of information and the training of health-care providers to use it must be improved if the use of pharmacogenetics is to become more common in clinical practice.

The health-care context

In the pharmaceuticals market, policy makers – defined broadly to include both public and private actors (*e.g.* managed care companies and other insurers) – generally articulate three objectives: efficiency, quality and access. Today, many OECD countries are particularly concerned with efficient expenditure. Since 2000, most have experienced rises in health-care expenditure that far outstrip the growth of GDP.

Across the OECD, spending on pharmaceuticals represents a sizeable – and increasing – proportion of health expenditure (Figure 5.1). The United States is the largest pharmaceuticals market, even though pharmaceuticals represent a relatively small percentage of the country's health-care budget. In the United States, spending for prescription drugs grew from slightly more than USD 40 billion in 1990 to more than USD 216 billion in 2006, an increase of about 540%.[1] Pharmaceutical sales in France and Japan are also large – approaching 20% of their respective health care budgets. While pharmaceutical expenditures in Australia and Germany are modest in comparison, both countries are experiencing considerable growth. This growth can be attributed, at least in part, to ageing populations and to increases in the incidence of cardiovascular and gastrointestinal diseases, diabetes, and diseases of the central nervous system.

Policy makers view growth in health-care spending with concern. Increasing health-care expenditures, shrinking funding for health care and growing consumer concerns about increasing co-payments have resulted in mounting pressure on drug companies to provide better drugs at lower prices. For the pharmaceutical industry, this means increased emphasis on cost-effective interventions and on the quality of the health care delivered. Quality is, however, an ambiguous term. No health-care system routinely measures broad patient outcomes. Impacts on health are often measured by narrow clinical endpoints rather than broad impacts on quality of life. In addition, publicly funded health-care systems tend to aim to provide equal access to health care for equal need. Like efficiency and quality, this goal is difficult to monitor and achieve. Under-utilisation of available cost-effective drug interventions is a common access problem. The ability to predict patient response to a given medicine more accurately and more quickly will be important for the pursuit of more efficient health expenditures and better quality and access to health care.

The role of pharmacogenetics in evidence-based medicine

Evidence-based medicine is the deliberate and appropriate use of the current best evidence in making decisions about the care of a patient (Sackett *et al.*, 1996). It brings together individual clinical expertise with external clinical evidence gathered from relevant research from basic science and especially from patient-centred clinical research. Evidence-based medicine enables clinicians to consider the accuracy and relevance of all appropriate diagnostic tests, the power of predictive markers and the safety and efficacy of potential therapeutic and preventive approaches. In addition, it provides a mechanism for moving from older, accepted clinical practices and treatments to new ones that deliver better results such as improved safety or effectiveness.

Pharmacogenetics is already making a significant contribution to evidence-based medicine. There are medicines on the market for which pharmacogenetics-based tests make it possible to differentiate among patient groups and provide essential guidance for the prescribing decisions of clinicians. In clinical practice, pharmacogenetic testing is done to: *i)* identify responders and non-responders to a treatment; *ii)* establish appropriate dosage; and *iii)* determine potential susceptibility to adverse reactions and, thus, to assess whether or not a particular treatment should be used (see Box 5.1).

Box 5.1. Pharmacogenetics in the clinic: different models

As the following examples demonstrate, pharmacogenetics can be applied in clinical settings to carry out testing for three different purposes:

- **To exclude patients from a treatment**. Abacavir® is an anti-viral medicine for which a small number (8%) of patients exhibit adverse reactions which are not dose-dependent and can be life-threatening. The factors that predict such adverse events are not yet well understood, but there is evidence of a genetic component. If a genetic basis could be established, a genetic test could be used to identify the patients susceptible to adverse events and exclude them from treatment.
- **To include patients in a treatment.** Herceptin® and BiDil® are examples of medicines for which a genetics-based test is used to predict the prospective efficacy of the therapy. In both cases, a positive test result is used to include the patient in the particular treatment regime. In the case of BiDil®, using this approach with Afro-Caribbeans raises a number of ethical issues, particularly related to the use of self-declared ethnicity as a surrogate for genotype.
- **To establish an appropriate dose**. Within the body, tricyclic anti-depressants are metabolised by cytochrome P450 (CYP450) enzymes. Genetic variations in these enzymes can lead to variations in the rate of metabolism. For some patients, the end result is adverse drug reactions; for others, the level of efficacy is poor. Monitoring blood-stream concentrations of these drugs is not practical. It would be valuable if specific variations in a patient's CYP450 genes could be used to predict the potential response in advance of choosing a drug and calculating dosage.

Treatment with Herceptin® is an example of identifying responders and non-responders. This medicine is a therapy for women with metastatic breast cancer whose tumours express too much of a protein known as HER2. Approximately one in four women with breast cancer is positive for HER2, a characteristic that is essentially a consequence of the individual's genetic make-up. Using HER2 as a genetic biomarker to test all women with breast cancer, it is possible to identify and treat those who will benefit from Herceptin®. This targeted use increases the observed effectiveness of Herceptin®. In effect, the percentage of women who will respond to the medicine is greatly enhanced within the targeted group, as compared to the percentage from the total population of women with breast cancer. This approach can change the risk-benefit ratio for treatment: as the medicine is prescribed only to women who are likely to benefit, so that the number of patients who experience side effects (or experience no therapeutic effect) is significantly reduced.[2]

BiDil®[3] is another example of a licensed medicine that is relevant for a restricted population. In 2005, the medicine was approved for use by the US FDA. As with Herceptin®, selection of the appropriate patient population clearly demonstrated the medicine's efficacy. In contrast to Herceptin® no clinical test is involved; self-identification of being black is the criterion used to identify patients with increased probability of benefiting from BiDil®. At a genetic level, the selection is based on increasing evidence that: *i)* genetic differences in ethnic populations may influence their response to different medicines (Lanfear *et al.*, 2004); and *ii)* ethnicity can be used as a surrogate for genetic markers to distinguish patient groups (Shah, 2004).

The use of such surrogate markers is not without some drawbacks, however, especially if there is genetic heterogeneity within self-identified communities and if there is an imperfect match between the marker, such as ethnicity or skin colour and the likelihood of a favourable clinical response to a drug.[4]

Pharmacogenetics can also be applied to established medicines. Many medicines have been approved for general use with little or no consideration of whether particular patients will respond to the standard therapeutic dose, or whether there is a need to predict which patients are more likely to be susceptible to unwanted side effects. There are non-responders to many established medicines – *i.e.* individuals who, for genetic reasons, will not benefit from a particular medicine (Ingleman-Sundberg, 2005). There are also individuals who experience adverse side effects when prescribed the standard therapeutic dose of a given medicine. Extensive evidence suggests that these responses are associated with the presence of particular versions of specific genes known to be involved in the metabolism of the medicine (Phillips *et al.*, 2001; Evans and Relling, 2004; Ingleman-Sundberg, 2004, 2005).

One particular family of genes and their proteins, the cytochrome P450 superfamily (CYP450), has been shown to be involved in the metabolism of approximately 25% of currently available medicines. Genetic variation in a number of the genes within this family, in particular CYP2D6, can cause patients to exhibit variable responses to commonly prescribed medicines. It is important to be able to identify in advance individuals who may be at risk of potentially life-threatening side effects. A recent pilot trial for the dosing of the drug Warfarin® demonstrated that an appropriate pharmacogenetic test for CYP2C9 variants can be used in a clinical setting. This test delivers benefits to the patient and the clinician by facilitating more rapid and more accurate prescribing decisions (Hillman *et al.*, 2005) (see Box 5.2).

Box 5.2. Pharmacogenetic testing for prescription of Warfarin® and for statins

Warfarin® is an established medicine for which adverse events are both significant and relatively frequent (Pirmohamed *et al.*, 2004). Research shows that it is possible to achieve a 62% success rate in predicting patient response to Warfarin® treatment. This can be achieved by testing patients for the presence of the CYP2C9 polymorphism and the vitamin K epoxide reductase polymorphism (VKORC1), and then combining these results with information about a patient's age and weight (Wadelius *et al.*, 2005). It is also possible to introduce such tests into clinical practice, thereby improving the quality of clinical care through more accurate dose prescription. This leads to fewer adverse events (Hillman *et al.*, 2005).

Statins are a class of hypolipidemic agents (*i.e.* lipid-lowering drugs). They are recognised as a highly effective treatment for lowering blood cholesterol levels and improving the health outcomes of patients at risk of heart disease. Their potential in other disease areas is being investigated, including diabetes and cancer. Researchers have identified 41 genes with potential relevance for response to statins. They have concluded that, in most cases, polymorphism of these genes will be relevant to the patient response to statins, as will specific other genetic variations that affect aspects of cholesterol metabolism (*e.g.* its absorption and production, as well as lipid catabolism) (Kajinami *et al.*, 2005). Given the number of genes potentially involved in response to statins, pharmacogenetic research is needed to determine who is most – and least – likely to benefit from statin treatment. Such findings could lead to improved and targeted management of coronary disease.

Very recently, the US FDA issued advice that it is useful to carry out genetic tests for CYP2C9 variants (as well as for variants in the VKORCI gene – encoding for the vitamin K epoxide reductase complex) before a physician prescribes Warfarin® (see Box 5.2).

The role of physicians

Physicians are important for developing demand for medicines and tests that improve the health care of patients. They are the primary prescribers of drugs, although other health professionals (such as prescribing nurses in the United Kingdom and the United States) also play a role. Pharmacists also have substantial pharmacological knowledge and experience, which can be used to improve the efficiency of prescribing (for example, by generic substitution).

In clinical practice, physicians use various parameters or clinical biomarkers to assist in diagnosing disease and in identifying the optimum treatment regime. Some biomarkers are readily measured in the physician's office (*e.g.* blood pressure and body mass index); others require access to specialist facilities and trained staff.

Box 5.3. Pharmacogenetic testing for thiopurine methyltransferase

One of the earliest tests for a genetic variation which was clinically important involved the enzyme thiopurine methyltransferase (TPMT). TPMT metabolises 6-mercaptopurine and azathioprine, two drugs used in a range of indications including childhood leukaemia and various autoimmune diseases.

In people with a deficiency in TPMT, metabolism must proceed by other pathways. One such pathway produces a metabolite that is toxic to bone marrow and the liver. The metabolite puts these people at risk of potentially fatal bone marrow suppression or hepatic failure. In 85-90% of affected people, this deficiency results from one of three variant alleles. One in 300 people have two variant alleles and need only 6-10% of the standard dose of the drug; if treated with the full dose, they will develop severe bone marrow suppression. For these individuals, the genotype predicts clinical outcome, a prerequisite for an effective pharmacogenetic test.

Around 10% of people (30 of the 300) are heterozygous (have two different alleles – one normal and one abnormal) and produce a smaller quantity of functional enzyme. Collectively, this group is at greater risk of adverse effects than patients with two variant alleles. Their genotypes are not necessarily predictive of clinical outcome, which makes it difficult to interpret a clinical test.

Clinical studies with thiopurine drugs show that in heterozygous patients an average of 78% of side effects are not associated with TPMT polymorphisms. Recent research on 6-mercaptopurine also shows that children who are heterozygous and are treated with equivalent doses to wild-type patients had a better response to treatment. This study shows how pharmacogenetics may contribute to refining and improving the treatment of patients with the most common genotypes.

The US FDA recently deliberated including a recommendation for TPMT deficiency testing in the packaging insert for 6-mercaptopurine and azathioprine. Previously, the insert information only carried a warning that inherited deficiency of the enzyme could increase the risk of severe bone marrow suppression. Now the insert will carry the recommendation that TPMT enzyme deficiency testing be considered in patients with clinical or laboratory evidence of severe toxicity, particularly bone marrow toxicity.

For the physician and the patient, it is the individual response that determines the selection of treatment regimes. Thus, physicians' interest in pharmacogenetic biomarkers will depend on their perception of the value of biomarkers in identifying individual response.

Several criteria must be met in order to encourage clinicians to include pharmacogenetic biomarkers in the range of tests they normally apply. First, the tests must be made available in a timely, reliable and recognised form, and must give clinically useful, reproducible results that can be readily interpreted and understood. Second, clinicians must be aware of and properly informed about both the benefits and limitations of the tests. Clearly, there is a need to educate physicians so that they remain up to date in a rapidly changing field.

It is not surprising to find that demand for pharmacogenetic testing by physicians and other clinical practitioners is currently limited. In general, such tests are used when it is known that they will enable accurate identification of patients who will benefit from a particular medicine (as with Herceptin®) or to avoid known side effects (as with TPMT, see Box 5.3). However, even with these drugs, there are great disparities in test usage and clinical uptake. Recent studies found that 84% of clinicians tested women with breast cancer for HER2 before prescription. In contrast, 53% of clinicians prescribed TPMT without any prior testing for TPMT deficiency (Enzing *et. al,* 2006).

The studies did note that this discrepancy in testing practice between HER2 and TPMT reflects several factors including: market size (*e.g.* the number of patients likely to need the treatment); whether the medicine is new or already in clinical practice; the costs and reimbursement of the treatment; and the degree to which physicians are concerned about possible liability. Above all, physicians must be informed and educated about test use and interpretation. Manufacturers played an important part in educating physicians about the added value of HER2 testing before prescription (Woelderink *et al.*, 2006).

Physicians and the challenge of integrating information

In the context of clinical care, the added value of pharmacogenetics ultimately derives from the capacity to transform basic science into information that supports clinical decision making. Integrating molecular tools in the practice of health care is of interest to physicians as long as it improves care decisions.

A more formidable challenge lies in implementing "point-of-care" data management tools. Physicians will want to be able to compare new approaches to traditional models of care, and to confirm therapy across a number of diagnostic biomarkers and platforms. In order to do so, they will need tools to capture and integrate environmental factors such as co-morbidity. They will also need to query significant amounts of patient data to gain insight into disease prevention, diagnosis and treatment. In short, physicians will need access to effective data query and data mining tools.

The complex task of pulling together the data required – and putting it into formats that both clinicians and researchers can use – remains daunting, particularly as no such systems or information-processing platforms currently exist. Successful, large-scale uptake of pharmacogenetics will depend on advances in medical information technologies. Indeed, as discussed in Chapter 3, the successful use of pharmacogenetics in drug development and design also will rely in part on these same advances.

Experts in the field have developed some basic informatics and computer science tools that can help with the transformation of medical record keeping. However, a great deal of informatics research is required to develop and deploy the architecture capable of integrating and managing all the data that will be needed at the point of care. This architecture will need to be based on standards for encoding and processing information.

Security of information is also critical. Within health-care settings, the aggregation and analysis of individual patient data, although coded, must remain identifiable to enable clinical care. Legal systems in most OECD countries currently provide guidelines for maintaining patient data confidentiality. Yet the sheer volume of data and the potentially large number of end users in clinical settings create the need for highly secure systems that incorporate sophisticated tools for authenticating authorised users and for monitoring access to patient data. A system needs to be developed with the capacity to provide output to specialised clinical decision-support tools and the flexibility to accommodate new data inputs as they arise. The information from pharmacogenetics needs to be integrated with other clinical information and data on environmental factors so that decision making is more effective and efficient and less risky. The system therefore needs to be understandable and useable at point of care.

Patient demand

Given the opportunity, patients and patient groups could be influential in driving the demand for pharmacogenetic tests and medicines. Patients are increasingly well informed about the genetic component of disease and the options for testing, diagnosis and treatment regimes. They are also aware of the availability of new medicines, including targeted drugs. Patients want the best medicine for their condition: their decision-making strategies include a willingness to consider the risk/benefit equation of a medicine.

Patients generally do not perceive pharmacogenetic testing and medicines as being any different from other medicines and techniques that may be applied. However, they do place high importance on a number of issues that follow from genetic testing, including concerns over privacy and confidentiality. Such issues are not associated with diagnosis and prescribing decisions, but they may affect patients' attitudes and, therefore, their overall interest in genetics-based medicines. Patient concerns relate directly to the process of obtaining genotypic information and cover topics such as the collection, storage, use and ownership of DNA samples and data and the nature of the consent required. Patients are also concerned about the end use of the data extracted from genotyping, particularly its potential release to third parties (*e.g.* employers, health insurers, etc.).

Some of these concerns can be addressed through legislation and other policy instruments;[5] others can be appropriately addressed through communication and by actively engaging patients and their families. In Sweden, for example, community-based support programmes for genetic diseases have shown that family engagement is beneficial to patients' health and significantly reduces the overall costs of managing patients (Olauson, 2005). These programmes emphasise the importance of communication between all parties and the patients; the aim is to ensure that patients are able to understand the issues and take responsibility for decisions.

Given the ready availability of medical information in the media, it is not uncommon these days for patients to be more informed than physicians, particularly primary care physicians, about specific treatments. The Internet raises the particular problem of direct-to-consumer marketing of genetic tests. In some jurisdictions a patient can have specific genetic tests performed on his/her sample, without seeking a physician's advice or obtaining a prescription. Unregulated tests, possibly of poor quality, are then performed for a patient who presents the results to a clinician with a request for intervention. This is generally regarded as an inappropriate route for the initiation of treatment decisions.

The issue of access

Access to medicines, both established and new compounds, is an important issue for patients. Several factors can influence how easily a patient or population group can access particular medicines. Some are pertinent to all health-care systems; others are an outcome of the divide in health care between the developed and developing economies.

The first factor that influences access is cost. Irrespective of the health-care system, it is necessary to meet the costs of tests and medicines. All OECD countries have introduced cost-sharing for pharmaceuticals as a means to ensure what they regard as appropriate usage and control demand. In many cases, patients may be required to pay a portion of these costs (which may be very high for premium-priced medicines), either directly or through some form of health insurance. End payers are price-sensitive and tend to seek the least expensive treatment. In some cases patients are excluded from treatment because of an inability to pay.

Costs are also an important element of reimbursement decisions by health insurers or other reimbursement systems. How assessments of health technology and cost-effectiveness may take account of pharmacogenetics is discussed in Chapter 6, along with how such assessments might be integrated into reimbursement systems and evaluations of overall health system performance. However, the point here is that exclusion from reimbursement systems – whether for genetic tests or treatment – provides an effective barrier to patient access.

Labelling practices may also raise access issues, which may or may not be linked to reimbursement. For example, labels on drugs may advise a physician to see the results of a genetic test before prescribing. However, many genetic tests are currently not reimbursed or are reimbursed at well below cost. This may mean that access to a genetic test (and the cost implications thereof) acts as to hinder access to a therapeutic even if the latter is fully reimbursed. Here, it is the reimbursement system rather than the label that is the block, but it serves to illustrate that labelling needs to take account of possible unintentional impacts.

Local factors may also influence access to a medicine or a test. There are variations in the knowledge and expertise of local physicians. In some countries, the rules applied by the reimbursement system also vary locally. In the United Kingdom, for example, a local factor has become known as "postcode prescribing"; because of decisions regarding the reimbursement system, the area in which a patient lives may determine the range of therapeutic options available for a given condition.

Access to pharmacogenetic medicines may be impeded if a patient resides in a developing economy (particularly, though not only, if they are poor). The problem is not simply financial resources, although these are relevant. Most clinical trials are carried out in Western countries and use Caucasian patients and thus take little account of how other populations' genetic backgrounds might alter patients' response to a medicine. Evidence shows, however, that patterns of genetic variation can be identified in specific ethnic groups and may affect on that group's response to a medicine (Goldstein *et al.*, 2003).

New approaches will be needed to extend trials to specific ethnic groups in developing countries. This has significant ethical and economic ramifications. A balance will have to be found between providing sufficient incentives to encourage such developments and dealing with their potential impacts on health-care systems, as this will require sophisticated data-recording, monitoring and control systems. These systems, as well as the price of new medicines, may be prohibitive for a developing economy, particularly when the need is substantial. Moreover, the lack of appropriate infrastructures can influence clinical and prescribing conditions.

Efforts are being made to address some of these issues. For example, one aim of the Pharmacogenetics for Every Nation Initiative (PgENI) is to establish local testing programmes (see Box 5.4 and also Box 2.1). Ensuring that pharmacogenetics benefits all nations will require inputs from regulators, governments, industry and patient organisations at the international level.

Box 5.4. Self-declared ethnicity as a surrogate for genotype: a problem of access

Evidence shows that patterns of genetic variation can be identified in specific ethnic groups and may affect that group's response to a medicine (Goldstein *et al.*, 2003). These findings are the basis of the Pharmacogenetics for Every Nation Initiative (PGENI), which seeks to identify patterns of.

These findings raise some interesting questions. For example, to what extent is self-declared ethnicity an accurate route to ensuring that patients are prescribed an effective medicine? This approach was used with BiDil® in a clinical trial in which skin colour was used as the marker for selecting patients for treatment. Is it possible to imagine a scenario in which using this "test" would actually lead to denial of access to the medicine on the basis of the patient's own assessment of his or her genotypic status?

…/…

Box 5.4. Self-declared ethnicity as a surrogate for genotype: a problem of access *(continued)*

A study of 300 Brazilians showed that among 200 self-declared white Brazilians and 100 self-declared black Brazilians, patterns of genetic variation revealed individuals deriving from European, African and Amerindian ancestry (Carvalho and Pena, 2005). This evidence of admixture suggests that apparent ethnicity is not an accurate tool for identifying patients for specific treatments. Furthermore, there is evidence that the genetic variation between ethnic groups is less pronounced than the genetic variation seen within an individual ethnic group, even when there is less historic mixing of genetic backgrounds than is seen in Brazil.

200 self-declared white Brazilians

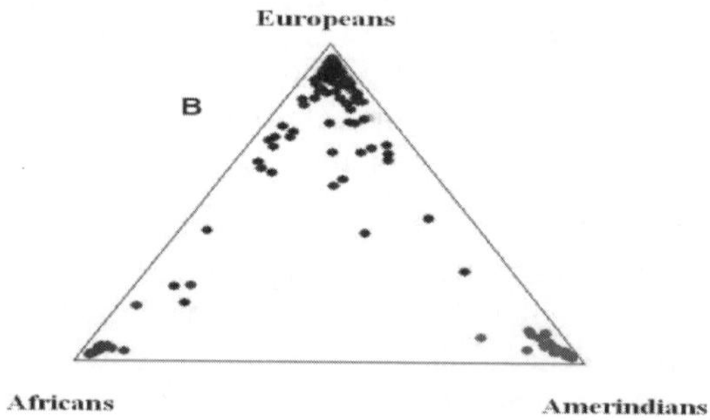

100 self-declared black Brazilians

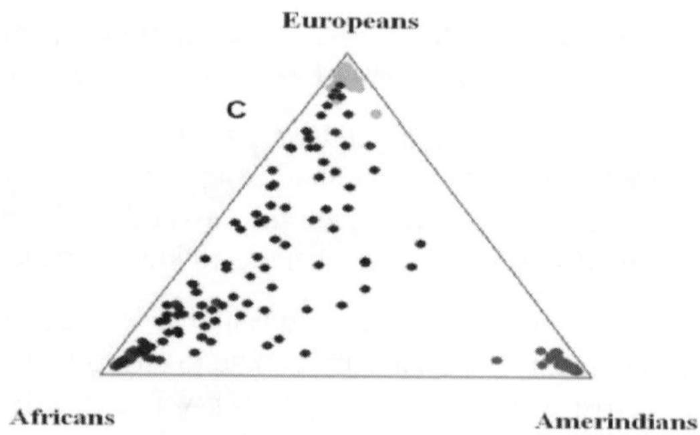

Source: Adapted from Suarez-Kurtz (2005), "Association Studies to Detect Drug-gene Interactions in Large Populations", presentation at the OECD workshop "An International Perspective on Pharmacogenetics: The Intersections between Innovation, Regulation and Health Delivery", Rome, Italy, 17-19 October, *www.oecd.org/sti/biotechnology/pharmacogenetics*.

Finally, whether in a developing or developed economy, guidance may be required on how the variable risks and benefits of treatment of individual patients will be handled by physicians and/or insurers in the pharmacogenetics age, as population polymorphisms and the variable responses they elicit lead to quite different estimations of risks and benefits for different individuals.

Health-care systems

At this stage, the extent to which pharmacogenetics will affect health care is not yet clear nor is whether its impact is likely to be revolutionary or evolutionary. A number of recent reports conclude that pharmacogenetics itself is unlikely to revolutionise health care in the near future. Whether or not this will prove to be the case (none of these reports has considered the role of pharmacogenetics in the light of the entire innovation cycle) pharmacogenetics should not be treated in isolation from other scientific and technological advances such as quantitative or digital real-time imaging, diagnostics based on high-throughput parallel processing, and automated systems. Taken together these could revolutionise, or at least transform, clinical care. The convergence of these tools is generating the data which will provide the potential to shift medical care from a reactive system (*i.e.* one that focuses on episodic treatment) to a proactive system that integrates knowledge platforms, predicts and manages risks, and ultimately provides not just targeted therapy but personalised medicine.

For example, the clinical practice of oncology has responded positively to the emergence of pharmacogenetic approaches. Moreover, it has been able to incorporate these novel approaches into clinical practice with minimal adjustments to existing systems. Some suggest that this implies the potential to realise the benefits of targeted medicines and biomarker tests without major modifications to current health-care systems. However, several factors account for the successful integration of pharmacogenetics in oncology. Cancers are known to be heterogeneous, to respond differently to pharmaceutical agents, and to require different treatment strategies. These characteristics are true of all types of cancer. Thus, therapeutic approaches have traditionally been flexible, targeted to the type of cancer, and responsive to complex decision-making tools to optimise health outcomes. In essence, cancer therapy has always required approaches that are similar in nature to pharmacogenetics. In addition, rapid decision making is vital for cancer patients: response to therapy is often a matter of life and death. Lastly, oncology is an evolving therapeutic area that has traditionally relied on strong links with clinical research.

The factors that have underpinned the success of pharmacogenetics in oncology may not be present in other clinical specialties. Moreover, it is possible that applying pharmacogenetics to other therapeutic areas will necessitate organisational changes in health-care systems and such changes may achieve more than the incremental adjustments that have occurred in oncology.

Education and workforce development

At present, several questions related to pharmacogenetics remain unresolved: Who will practice pharmacogenetic medicine? In what settings? And under what care structures? Current applications suggest that the delivery of pharmacogenetic medicines and testing may require broader involvement in patient care by health-care professionals other than the general practitioner or clinical specialists and in particular, by pharmacists and nurses. Pharmacogenetics may prompt a new division of labour: for example, physicians might be responsible for prescribing a drug, after which pharmacists or nurses might use the patient's genetic information to determine the correct sub-type of the drug and the appropriate dosage. This scenario would require greater emphasis on genetics and pharmacogenetics in training programmes (both undergraduate and postgraduate) for physicians, nurses and pharmacists. Regulators, payers, managers of health-care systems and others involved in the delivery of health care would also need a basic understanding of the nature and implications of genetics, that is, a knowledge level sufficient to grasp the impact of genetics on health-care systems and clinical practice (Gurwitz, 2005).

Successful integration of pharmacogenetics will also depend on appropriately informed patients. Safety of medicines and health-care costs are regular discussion points in all societies. There is growing appreciation of the reality that not all medicines work for everyone. Public consultations show a general understanding of the potential benefits associated with the development of pharmacogenetics-based testing.

Currently, much of the awareness of – and education about – the safety and efficacy of medicines is acquired by individuals through the media and Internet, often in response to a personal need. While self-education is to be encouraged, experience suggests that governments and other bodies could benefit from and enhance the quality of public discourse on this issue through more formal educational and health-care promotion initiatives.

Conclusions

Medicines for which pharmacogenetics-based tests enable differentiation between patient groups are now on the market and provide essential guidance for the prescribing decisions of clinicians. Use of these tests is limited but growing (the August 2007 US FDA decision on the utility of genetic tests prior to prescribing Warfarin® may prove to be something of a landmark). The perceptions and behaviour of physicians and patients will be decisive for the uptake and diffusion of pharmacogenetics. Successful, large-scale uptake of pharmacogenetics by clinicians will essentially depend on:

- evidence about the clinical utility of pharmacogenetic testing;
- the ability to integrate information across platforms and communicate across multiple end users within a health-care setting;
- availability of pharmacogenetics and other relevant information at the point of care.

For the most part, patients do not distinguish between pharmacogenetic drugs and testing and more traditional medicines and techniques. However, they are concerned about ethical and legal issues related to the handling of samples and data, the nature of the consent required, and the end use of the data extracted from genotyping and its potential release to third parties. The way these issues are addressed may affect patient attitudes toward genetics-based medicines.

Several factors may also influence how easily a patient or population group accesses particular pharmacogenetic medicines. Some are pertinent to all health-care systems; others are more specific to pharmacogenetics and reflect the divide that exists between health care in developed and developing economies. Evidence shows that patterns of genetic variation can be identified in specific sub-populations and ethnic groups and may affect that group's response to a medicine. Supportive approaches will be needed to extend pharmacogenetics research to these specific ethnic groups, particularly for well-established essential medicines.

At present, several questions related to pharmacogenetics remain unresolved. Who will practice pharmacogenetic medicine? Where? Under what care structures? With what sorts of effects on the organisation of changes in health care systems? Current applications suggest that pharmacogenetics is unlikely to revolutionise health care in the near future. It is, however, generating a body of data that will help transform medical care from a reactive system (*i.e.* one focused on episodic treatment) to a proactive system that can provide personalised medicine. The best example of this proactive and increasingly personal approach to clinical care is in oncology, which has been quick to adopt pharmacogenetic medicine.

Notes

1. Figures from the US Centers for Medicare and Medicaid Services.
2. See *www.herceptin.com/herceptin/patient/index.jsp*.
3. BiDil®: isosorbide dinitrate/hydralazine hydrochloride.
4. The use of surrogate markers for genetic polymorphisms is considered further in the Annotations to the *OECD Guidelines for Quality Assurance in Molecular Genetic Testing*.
5. Three OECD instruments address elements of such issues, namely: *Guidelines for Quality Assurance in Molecular Genetic Testing* (2007); *OECD Best Practice Guidelines on Biological Resource Centres* (2007); and ongoing work towards guidelines on human genetic research databases (expected 2009). For more information see *www.oecd.org/sti/biotechnology/ genomics.*

References

Centers for Medicare and Medicaid Services (2004), *Medicare National Coverage Determinants (NCD) Coding Policy Manual and Change Report*, Centers for Medicare and Medicaid Services, Baltimore, MD.

Carvalho C.M.B. and S.D.J Pena, (2005), Optimization of a Multiplex Minisequencing Protocol for Population Studies and Medical Genetics. *Genetics and Molecular Research,* 4, pp. 115-125

Evans, W.E. and M.V. Relling, 2004, "Moving Towards Individualized Medicine with Pharmacogenomics" *Nature,* 429, pp. 464-468.

Enzing, C. et. al. (2006), "Case Studies and Cost-Benefit Analysis of HER2 and TPMT in Four EU Member States," European Commission. as accessed on 9 April 2009: *ftp://ftp.jrc.es/pub/EURdoc/eur22214wp2.pdf.*

Goldstein, D.B., S.K. Tate, and S.M. Sisodiya (2003), "Pharmacogenetics Goes Genomic", *Nature Reviews Genetics,* 4, pp. 937-947.

Gurwitz, D. (2005), "Are Health Care Systems Ready to Deliver Pharmacogenetics as Standard of Care? Predicting the Needs and Setting the Strategies", paper presented at OECD workshop "An International Perspective on Pharmacogenetics: The Intersections between Innovation, Regulation and Health Delivery", Rome, 17-19 October, *www.oecd.org/sti/biotechnology/pharmacogenetics.*

Hillman, M.A., R.A. Wilkie, S.H. Yale, H.J. Vidaillet, M.D. Caldwell, I. Glurich, R.L. Berg, J. Schmelzer and J. K. Burmester (2005), "A Prospective, Randomized Pilot Trial of Model-Based Warfarin Dose Initiation using CYP2C9 Genotype and Clinical Data", *Clinical Medecine and Research,* 3, pp. 137-145.

Ingelman-Sundberg, M. (2004), "Pharmcogenetics of Cytochrome p450 and Its Applications in Drug Therapy: the Past, Present and Future Trends", *Pharmacological Sciences,* 25, 193-200.

Ingelman-Sundberg, M. (2005), "Genetic Polyamorphisms of Cytochrome p450 2D6 (CYP2D6): Clinical Consequences, Evolutionary Aspects and Functional Diversity", *The Pharmacogenomics Journal,* 5, pp. 6-13.

Lanfear, D.E., S. Marsh, S. Cresci, W.D. Shannon, J.A. Spertus, and H.L. McLeod (2004), "Genotypes Associated with Myocardial Infarction Risk Are More Common in African Americans than in European Americans", *Journal of the American College of Cardiology*, no. 44, 165-167.

OECD (2007), *OECD Best Practice Guidelines on Biological Resource Centres*, OECD, Paris.

Olauson, A. (2005), "Impacts on Human Health and Health Care Systems", paper presented at OECD workshop "An International Perspective on Pharmacogenetics: The Intersections between Innovation, Regulation and Health Delivery", Rome, 17-19 October, *www.oecd.org/sti/biotechnology/pharmacogenetics.*

Phillips, K.A., D.L. Veenstra, E. Oren, J.K. Kee and W. Sadee (2001), "Potential Role of Pharmacogenomics in Reducing Adverse Drug Reactions", *Journal of the American Medical Association,* 286, no. 18.

Pirmohamed, M., S. James, S. Meakin, C. Green, A.K. Scott, T.J. Walley, K. Farrar, B.K. Park and A.M. Breckenridge (2004), "Adverse Drug Reactions as Cause of Admission to Hospital: Prospective Analysis of 18,820 Patients", *British Medical Journal* 329, pp. 15-19.

Sackett, D.L., W.M.C. Rosenberg, J.A.M. Gray, B. Haynes and W.S. Richardson (1996), "Evidence-based Medicine: What It Is and What It Isn't", *British Medical Journal*, 312, pp. 71-72.

Shah, R.R. (2004), "Drug Development and Use in the Elderly: Search for the Right Dose and Dosing Regimen", *British Journal of Clinical Pharmacology*, 58, p. 452.

Suarez-Kurtz, G. (2005), "Association Studies to Detect Drug-gene Interactions in Large Populations", presentation at OECD workshop "An International Perspective on Pharmacogenetics: The Intersections between Innovation, Regulation and Health Delivery", Rome, Italy, 17-19 October, *www.oecd.org/sti/biotechnology/pharmacogenetics*.

Wadelius, M., L.Y. Chen, N. Eriksson, J. Ghori, C. Wadelius, D. Bentley, R. McGinnis and P. Deloukas, (2005), "Common VKORC1 and GGCX Polymorphisms Associated with Warfarin Dose", *Pharmacogenomics,* 5(4), 262-70.

Woelderink, A. *et al.* (2006), "The Current Clinical Practice of Pharmacogenetic Testing in Europe: TPMT and HER2 as Case Studies", *The Pharmacogenomics Journal* 6, pp. 3-7.

Chapter 6

Regulatory Authorities and Reimbursement Mechanisms

This chapter looks at the role of regulators and reimbursement systems in facilitating the uptake of pharmacogenetics by health-care systems. Regulators and coverage policies will be strong factors in the rate of diffusion of pharmacogenetics. Reimbursement policies and cost-containment strategies for drug and test-use policies and clinical guidelines will be very important. The challenges for payers and the trade-offs they may have to make are also considered.

Both regulatory and reimbursement systems will play a key role in creating a context that encourages industry and the wider health-care community to take up pharmacogenetics. Regulators are in a position to significantly improve the introduction of pharmacogenetics – for both new and established medicines – as a product of the health-care industry and a tool in clinical care. This power to influence could be further increased through effective development of transnational regulatory approaches and policies and concerted efforts to bridge the gaps in relevant practices. Strong interaction and co-ordination between national regulatory authorities will be vital to fulfilling their role as "enablers" for the application of pharmacogenetics.

Strategies to reward increased therapeutic benefit through reimbursement decisions will also influence the rate of diffusion of pharmacogenetics. Two aspects will be of particular importance: reimbursement policies for drugs and their associated tests, and economic evaluation and evidence-based policies. The chapter concludes with a discussion of the challenges for payers and the relative trade-offs.

The regulatory authorities

A central objective for all regulatory authorities is to regulate the introduction of new medicines so as to ensure their quality, safety and efficacy. Regulators view pharmacogenetics as a useful tool to reduce uncertainties in drug therapy, particularly those related to efficacy and potentially adverse side effects. Thus, they actively encourage this novel approach.

Regulators also have other reasons to support pharmacogenetics. In recent years, they have become increasingly concerned about the falling numbers of new drugs that reach the market and about the eventual fate of those that do. Evidence shows a rise in the post-marketing withdrawal of effective medicines, typically as a consequence of adverse drug reactions (ADRs). Regulators also keep close watch on the levels of morbidity and mortality associated with ADRs and on the impact of insufficiently efficacious medicines. In fact, these concerns are equally relevant to both new and established medicines (Shah, 2006; Shah, 2004). The primary aim of regulation is to protect the patient. However, the loss of medicines that may be efficacious in a clearly defined (or at least definable) population is evidence of an approach that errs on the side of risk avoidance and is suboptimal in terms of the efficiency of delivery of health benefits.

Exactly how regulatory authorities might intervene to create a context that supports pharmacogenetics will be influenced by whether the therapies in question are new, in development or on the market. Different strategies might be required for drugs currently available, for those that are still covered by patent (and therefore of significant concern to the patent holder) and for established medicines now manufactured as generics.

New medicines

Regulatory authorities are keenly aware of the potential opportunities pharmacogenetics offers for the development of new medicines. A recent trend among regulatory authorities worldwide is to demand that pharmaceutical companies provide clear indications of which patient populations are likely to benefit from a new treatment. This requirement encourages pharmacogenetic research but may drive up costs – at least in the short term. Regulatory requirements to demonstrate improved efficacy and cost-effectiveness of drugs also prompt industry to provide more pharmacogenetic data (Rägo, 2003; Dickson *et al.*, 2003).

A number of regulatory authorities have issued guidance on the use of pharmacogenetic data. For example, in 2004, the US Food and Drug Administration (US FDA) established a new process of consultation with industry, known as Voluntary Genomic Data Submission (VGDS) (US FDA, 2005). Through this process, the US FDA invites submission of exploratory pharmacogenomic data on drugs or candidate drugs. This early exchange of data enables regulators to gain a better understanding of how pharmaceutical companies are applying genomics in drug development – and to identify questions and issues of significant scientific uncertainty. The US FDA is committed to this consultation process as part of a broader aim to identify the unknown issues and questions that might impede the uptake of genomics. In 2005, the US FDA released a preliminary concept paper concerning co-developed drug-device combinations which examined regulatory processes as well as related analytical and clinical issues.

Within the European Community (EC), the European Medicines Agency (EMEA) has similar initiatives. In particular, the EMEA has begun informal briefing meetings with industry, according to guidelines released to industry in 2004 and released to the public for further consultation in 2005. Following the signing (in the second quarter of 2005) of bilateral confidentiality arrangements, the EC EMEA and the US FDA have organised joint briefing meetings and discussions of VGDS.

In 2001, the Japanese Koseirodosho issued two guidance documents for pharmacogenetics: *Clinical pharmacokinetics/pharmacodynamics* (ELD Notification No. 796: 1 June 2001); and *Drug-drug interaction Study* (ELD Notification No. 813: 4 June 2001). In March 2005, the Japan Koseirodosho published guidelines entitlted: *Guidance on Information Submissions for Pharmacogenetic Clinical Trial Guideline* (ELD Notification No. 0318001:

18 March 2005). This guideline invites sponsors to voluntarily submit information regarding the purpose of trials, the target population, testing methods, and the number of subjects.

In Canada, Health Canada released a guidance document, *Submission Pharmacogenomic Information* in February 2007 following broad stakeholder consultation.

These and other initiatives clearly signal that regulatory authorities in leading pharmaceutical markets are proactively engaging companies to support the application of pharmacogenetic approaches in drug discovery and development. However, significant regulatory hurdles remain, at least in some countries or regions.

Clinical research and clinical trials

In the context of pharmacogenetics, a policy question of great relevance for regulators is how to improve the environment for clinical research and clinical trials so as to enable progress. In the United Kingdom, for example, the Clinical Research Collaboration is a major new initiative that aims to support innovation through new investments in infrastructure to underpin clinical research and to streamline regulatory procedures.[1] In the United States, the National Institutes of Health (NIH) recently launched a similar initiative, Re-engineering the Clinical Research Enterprise.[2] At the heart of these initiatives is the idea that, to be successful, clinical research needs a broad platform of partnerships – particularly among organised patient communities, community-based health-care providers and academic researchers. This networked approach to the innovation cycle, or the "clinical research enterprise" as the NIH puts it, is attracting more and more support as a concept from stakeholders, and is discussed in Chapter 3.

These initiatives also recognise that regulatory authorities can make a significant contribution by establishing clear regulatory pathways. Indeed, stakeholders agree that for pharmacogenetics to move forward, clear regulations are needed in areas such as patient recruitment, informed consent, sample collection and storage. To date, regulatory requirements for these activities are generally less developed than those that define drug approval procedures. The pharmaceutical industry remains constrained by regulatory bottlenecks such as lengthy approval times for research projects involving patients (in some cases because research ethics boards are not well informed) and non-uniformity of requirements for sample collection and storage across countries – or indeed, even within countries. Despite the existence of a significant body of work reviewing these issues in the context of human genetics research and biobanks, progress toward harmonising requirements has been slow.3 Some progress is being made, however, at least at the OECD level, following the publication of reports providing internally agreed *Guidance on Best Practices for Quality Control of Human-derived Biological Resources* (OECD, 2007) and on *Creation and Governance of Human Genetic Research Databases* (OECD, 2006).

In addition, mechanisms to promote the process of biomarker validation could also significantly influence progress. Systematic international efforts are needed to improve the biomarker validation process.

When evaluating new medicines, regulatory authorities will want to balance benefits and needs, particularly the benefits of supporting innovation and change to enable the introduction of new, pharmacogenetics-based medicines and the need for careful and accurate decision making. For pharmaceutical firms, timing of market entry and market uptake can be of critical importance to the financial viability of new products. If regulators are reluctant to make a "yes" or "maybe" decision owing to a lack of good evidence, the consequences may be significant for the sustainability of innovators and for the supply of innovative products. Early dialogue between innovators and decision makers may help to reduce the incidence of negative decisions, which can sometimes be based on lack of understanding or uncertainty in the evidence base. Decreasing the odds that the regulator will arrive at such a decision would help minimise the innovator's investment risk. Current evidence shows that by establishing an iterative dialogue with stakeholders (industry, clinicians, etc.) while a new drug is under development (see Figure 6.1), regulatory authorities can positively influence the introduction and application of pharmacogenetics (see also OECD, 2005a, Figure 8.1).

One way to stimulate this dialogue is through conditional marketing approval, which is sometimes granted in the case of serious diseases. Conditional marketing approval recognises the promising nature of the clinical evidence in patients, as well as the need for further follow-up to verify the clinical benefits. Figure 6.1 shows the iterative pattern of interaction between regulatory authorities, the pharmaceutical industry and clinical users of a medicine during its development and after conditional marketing approval. Every positive decision for conditional marketing approval carries an obligation for careful post-authorisation surveillance to evaluate whether the claims for therapeutic usage, efficacy and safety are validated in routine clinical practice. The aim is to encourage evidence-based clinical practice, while allowing for innovation and accelerated access to new medicinal products that address unmet needs.

Figure 6.1. Stakeholder interaction during drug development and after conditional marketing approval

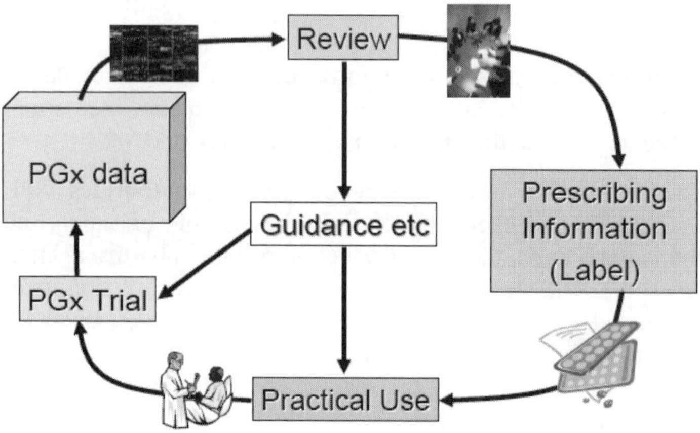

Source: Uyama (2005), "Perspective and Strategy on Pharmacogenetics", presentation at the OECD workshop "An International Perspective on Pharmacogenetics: The Intersections between Innovation, Regulation and Health Delivery", Rome, October, *www.oecd.org/sti/biotechnology/pharmacogenetics*.

Two criteria must be met in order to make conditional approval an effective means of assessing new and emerging technologies (OECD, 2005a). First, it is essential to gather more data. This requires not only resources but also a commitment from stakeholders: all parties must agree on a procedure to collect a minimum data set that will provide additional data regarding the key decision-making criteria. Second, conditional approval can only work if decision makers can, upon receiving new information, re-evaluate their decision – not just in theory, but also in practice and under realistic conditions. Stakeholders must agree to the decision-making process and accept the final outcome. Conditional approval processes may also be enhanced through risk-sharing agreements between the sponsors of a drug or technology and public authorities.

A conditional approval mechanism enables early access to a medicinal product or technology. At the same time, it recognises uncertainty in the evidence base, which may provide impetus for incorporating risk-sharing mechanisms. Ultimately, such mechanisms allow decision makers to account for the value of innovation, based on real-life data. Conditional approval mechanisms offer significant prospects for improving the integration of pharmacogenetics into health care. A review of best practices from this or similar initiatives may facilitate pharmacogenetics clinical research and uptake across OECD countries.

Established medicines

Pharmaceutical and devices industries have weak incentives to apply pharmacogenetics to established medicines, despite the fact that doing so would benefit patients and society as a whole by reducing the number of adverse drug reactions. New or unexpected serious adverse drug reactions are often only observed once a medicine has been on the market for some time and many thousands of patients have been exposed to it. Thus, the investigation and application of pharmacogenetics to established medicines becomes an issue of public policy.

Available data reveal that a significant percentage of drugs are withdrawn during the first five to ten years following their introduction to the market. For example, a study of drugs approved for marketing in Spain, the United Kingdom and the United States between 1974 and 1993 shows that 3-4% were subsequently withdrawn for safety reasons (Bakke *et al.,* 1995). This is not altogether surprising. Regulatory authorities approve drugs on the basis of population data on efficacy and safety from clinical trials, which are extrapolated to the entire population on the broad assumptions that the trial population is representative of all patients and that the ultimate patient population is a homogenous group.

A number of established medicines, which are now well beyond patent protection, may cause significant adverse reactions in some patients. These include medicines such as aspirin, ibuprofen and other common drugs such as anti-depressants, beta-blockers and opiates (Pirmohamed *et al.,* 2004). Cost-benefit analysis suggests that, for many of these medicines, developing specific tests to be used prior to prescription offers little or no benefit. However, in some circumstances, there would be significant benefit from using tests to aid in diagnosis and prescription. This is particularly true when the disorder being treated is serious or when a test is highly predictive and can lead to a modified therapeutic intervention. Prior testing can also be useful when monitoring the drug to adjust the dose is impractical or ineffective, when simpler therapeutic alternatives are unavailable, or when high costs are associated with the actual treatment or with failure to treat.

The issue, then, is how best to identify, validate and demonstrate the clinical utility of biomarkers which can be used to predict patient response to these medicines. Major prospective studies are needed to identify relevant genetic markers for patient stratification and to advance understanding of the application of pharmacogenetics to these common drugs, including the benefits for clinical practice.[4] Prospective studies would allow regulatory authorities to determine whether it is necessary to modify the labelling of specific established medicines in order to improve clinical practice and patient health. However, as the cost of such trials may be very high, one strategy might be to form public-private partnerships to carry them out nationally or possibly internationally.

An alternative approach for established medicines is to use existing prior knowledge. The risks of such compounds tend to be well known; in addition, mechanisms to identify at-risk patients are already available. Coupled with clinical experience, the known factors make it possible to develop a mechanistic hypothesis for adverse reactions which establishes associations between cause and effect, such as would be seen with a medicine for which there is a dose-response. Other data available – including case reports, case series and cross-sectional studies – may provide some descriptive evidence. These data could provide a basis for using case-control studies, which can be undertaken relatively quickly and at lower cost. For example, research might be conducted to compare genotypes of patients showing different responses to the medicine of interest.

Labelling of pharmacogenetic drugs

Regulatory authorities in OECD countries have recently launched systematic reviews of marketed drugs to decide whether the labelling should recommend genotyping to guide therapy. Pharmacogenomic information is contained in about 10% of labels for drugs approved by the US FDA. The number of labels containing such information has increased significantly over the last decade. Most drug labels, however, provide no immediate recommendation for a specific action (*i.e.* genetic testing); only a few recommend or require genetic testing and specify the use of these markers for reaching a therapeutic decision (see also the discussion of access issues in Chapter 5).

The US FDA published in 2006 a list of valid genomic biomarkers for FDA-approved drug labels.[5] The list gives comprehensive information on these markers and links to pharmacogenomic data, taking into account the regulatory contexts in which these biomarkers were approved. Genomic biomarkers can play an important role in identifying responders and non-responders, avoiding toxicity and adjusting dosage to optimise efficacy and safety. For drug labels, these genomic biomarkers have been classified by the US FDA on the basis of their specific use, for example:

- clinical response and differentiation,
- risk identification,
- dose selection guidance,

- susceptibility, resistance and differential disease diagnosis,
- polymorphic drug targets.

In 2006, a US FDA committee reviewed the evidence available on ADRs related to Warfarin® and launched a programme of prospective studies to validate a mathematical model (algorithm) that would predict dosing and incidence of adverse reactions and guide prescriptions.[6] In the United Kingdom, a prospective trial of 2 000 patients treated with Warfarin® is being funded by the Department of Health (Pirmohamed *et al.,* 2004). In August 2007, the US FDA issued guidance on the utility of genetic testing as a predictor of patient response to Warfarin® (see the section on pharmacogenetics in evidence-based medicine in Chapter 5).

Linking pharmacovigilance with pharmacogenetics

Pharmacovigilance refers to efforts undertaken to detect, evaluate and prevent undesirable side effects of a medicine after it receives marketing approval. Pharmaceutical products undergo extensive testing and review by means of controlled clinical trials prior to marketing. However, even when a drug passes Phase III trials, these usually involve no more than 1 000-2 000 patients and last three or four years. Thus, Phase III clinical trials often fail to detect rare side effects or side effects associated with long-term administration of a drug. In addition, most patients enrolled in clinical trials have relatively uncomplicated diseases and are drawn from select populations that meet rigorous inclusion and exclusion criteria. For these reasons, pre-marketing data often do not apply to a more "realistic" population that includes the elderly, pregnant women, children, and patients with more than one disease (who require treatment with multiple drugs). Such patients are among those likely to be exposed to a drug after marketing begins. Pharmacovigilance is crucial for providing additional safety information that cannot realistically be collected prior to drug approval. It can help identify risk factors, reveal interactions with other medicinal products, and support the analysis and provision of information on the safe and effective use and regulation of medicinal products.

In the context of pharmacogenetics, pharmacovigilance is extremely important. The documentation of adverse drug reactions offers the opportunity to further study possible genetic predictors of adverse drug reactions. Such research could lead to the development of new pharmacogenetic tests of demonstrated clinical value. It could also help regulators and other relevant organisations assess the cost-effectiveness of pharmacogenetics.

ADR reporting by health-care professionals is the most fundamental source of information on the safety of a medicinal product. It was conceived in the 1970s in direct response to the failure to recognise thalidomide phocomelia – the physical defects of "thalidomide babies" – until the drug had affected approximately 10 000 foetuses. At the time, it became evident that a notification system could lead to earlier recognition of the association between a specific drug and a very rare disorder. Almost all OECD countries now have national pharmacovigilance guidelines and an infrastructure for reporting ADRs. Under the auspices of the International Conference on Harmonisation (ICH), pharmacovigilance has acquired high priority and momentum at the international level.

Linking ADR reporting to pharmacogenetic studies could provide an effective means of rapidly developing the pharmacogenetics knowledge base and help to identify factors that increase the risk of unwanted outcomes from drug therapy. It will also help to establish the circumstances under which genotyping should be performed prior to starting drug treatment and to tailor drug treatment for individual patients. To achieve these goals requires a methodology that enables systematic estimation of how specific genetic factors confer risk of susceptibility to specific ADRs. Such a methodology would have to be acceptable to the general population. At the same time, efforts to link pharmacovigilance systems to pharmacogenetics will depend on international agreement and co-operation, the establishment of appropriate incentives, and broad acceptance on the part of physicians and patients.

The impact of reimbursement systems on pharmacogenetics

Reimbursement systems for medicines, devices and clinical treatments vary throughout the world and address the issues of the costs of and payment for medicines in different ways. These systems, whether public or private, reflect underlying national attitudes towards provision of care. All reimbursement systems have established a number of measures to manage, contain or regulate the provision of medicines. It is not within the scope of this report to review these systems and measures in detail, but rather to highlight features that may affect the uptake and diffusion of pharmacogenetics.

In flexible systems, reimbursement decisions can be used to send signals to industry about the price levels considered acceptable for specific clinical gain. In an ideal world, reimbursement systems should respond positively to the increased therapeutic benefit of a drug: this would effectively offer a "reward" for true innovation and gains in clinical effectiveness. At present, however, few pricing schemes are founded on increased therapeutic benefit. The use of economic evaluation and evidence-based policies for drug coverage could go some way towards achieving pricing that is more reflective of benefits.

Evidence-based coverage policies

Several OECD member countries have established a variety of frameworks and systems to assess new technologies and pharmaceuticals using an evidence-based approach (see Box 6.1). The OECD recently reviewed a number of these systems, which are highlighted below (OECD, 2005a).

Box 6.1. Frameworks for evidence-based review of new health technologies

Agency for Healthcare Research and Quality (AHRQ)
www.ahrq.gov

The AHRQ is the health services research arm of the US Department of Health and Human Services (HHS), complementing the biomedical research mission of its sister agency, the National Institutes of Health. The AHRQ funds research on health-care quality, costs and outcomes, and on patient safety. As a part of its function, the AHRQ funds evidence-based practice centres in Canada and the United States which specialise in the methods and conduct of systematic reviews.

Canadian Agency for Drugs & Technologies in Health (CADTH)
http://cadth.ca/

The primary mission of the CADTH (formerly known as the Canadian Co-ordinating Office for Health Technology Assessment or CCOHTA) is to provide timely, relevant, rigorously derived, evidence-based information about drugs and other health technologies. It also supports the decision-making process in Canada. Federal, provincial and territorial health-care decision makers rely on the CADTH to provide credible information and impartial advice. The CADTH's mandate is distinct from that of regulators, which typically focuses on determining which drugs or technologies should be used to achieve the best outcomes in terms of both patient health and effective operation of the health-care system.

The Cochrane Collaboration
www.cochrane.org

The Cochrane Collaboration is an international collaboration of researchers, brought together for the purpose of conducting and disseminating the results of systematic reviews. This group also conducts research in the methodology of systematic reviewing.

Drug Effectiveness Review Program (DERP)
www.ohsu.edu/drugeffectiveness

DERP is a collaboration of US-based public and private organisations (including 15 states) which have joined together to provide systematic evidence-based reviews of the comparative effectiveness and safety of drugs. DERP focuses mainly on widely used drug classes and seeks to apply its findings to inform public policy and related activities.

…/…

Box 6.1. Frameworks for evidence-based review of new health technologies *(continued)*

Effective Health Care
www.effectivehealthcare.ahrq.gov

Effective Health Care is a section of the US Medicare Modernization Act of 2003 (MMA). The programme focuses on effectiveness and improved health outcomes. It also authorises the Agency for Healthcare Research and Quality (AHRQ) to conduct research, demonstrations and evaluations designed to improve the quality, effectiveness and efficiency of Medicare, Medicaid and the State Children's Health Insurance Program (SCHIP). The programme emphasises a transparent process with public input. Its goals are two-fold: *i)* to develop valid evidence about the comparative effectiveness of different treatments and appropriate clinical approaches to difficult health problems; and *ii)* to make the information easily accessible and understandable to decision makers.

National Co-ordinating Centre for Health Technology Assessment (NCCHTA)
www.ncchta.org

The NCCHTA is an organisation funded by the UK Department of Health to provide health technology assessment to the National Health Service (NHS). As a part of this research, the NCCHTA commissions the undertaking of broad-scale systematic reviews which cover a wide range of health-care topics, including systematic review methods.

National Institute for Clinical Excellence (NICE)
www.nice.org.uk/

NICE is part of the UK National Health Service (NHS). Its role is to provide patients, health professionals and the public with authoritative, robust and reliable guidance on current "best practice". As a part of its remit, NICE commissions technology assessments which include both systematic reviews of clinical evidence and economic evaluations of interventions, such as drugs.

Pharmaceutical Benefits Advisory Committee (PBAC)
www.health.gov.au

PBAC is an independent statutory body (established in Australia on 12 May 1954 under section 101 of the National Health Act 1953) with a mandate to make recommendations and provide advice about which drugs and medicinal preparations should be made available as pharmaceutical benefits. No new drug may be made available as a pharmaceutical benefit unless the Committee has so recommended. In December 1993 (under section 101A of the National Health Act), PBAC established the Economic Subcommittee. Its role is to: *i)* review and interpret economic analyses of drugs submitted to the PBAC; and *ii)* advise the PBAC on these analyses, and on technical aspects of requiring and using economic evaluations.

Agencies and organisations charged with conducting evaluations of health economics follow a recognised process of identifying the issues and topics for assessment, gathering and interpreting evidence pertaining to the problem, and drawing conclusions and recommendations, which are then disseminated. In general, policy makers want to know whether new diagnostics and/or treatments are substitutes for existing methods or are "add-ons". They also want to know whether the new alternatives are cost-effective, cost-reducing or cost-neutral, and who the target population may be. The most commonly used technique is to assess the cost-effectiveness of a new technology by comparing the costs and outcomes of alternative treatments and then calculating the incremental cost-effectiveness ratio (ICER; see Box 4.1, Chapter 4, for further explanation).

For genetic tests, evidence-based decisions to provide coverage involve several factors. First, existing data on a test's analytical and clinical validity must be assessed; when the necessary data are available, assessment may also be based on evidence of clinical utility in relation to the populations to which the test is applied. Making a decision on coverage may also involve evaluation of the test's costs, cost-effectiveness and impact on health outcomes in relation to established alternatives as well as the subsequent costs of responding to the test results and patient preferences.

Reimbursement of genetic testing

A number of measures aim to contain or regulate the provision of genetic testing, including whether this testing can be provided by both the public and private sectors. Unless new and more appropriate schemes are developed, it is likely that the same provisions will apply to pharmacogenetic testing.

In most OECD member countries, reimbursement is not necessarily based on value added. The average public laboratory receives a standard budget for each sample examined, independent of the type of sample, the test or the work required to investigate it. This practice may discourage low-volume, technically complex and expensive testing procedures. It may also constrain the development and diffusion of new genetic tests and drive the centralisation of testing services.

Public and private health insurers play a significant role in defining patient and provider use and access to genetic tests. In most OECD countries, reimbursement of genetic tests by public insurance is conditional on medical referral. In many countries, reimbursement is also contingent on the tests being carried out by an officially recognised genetic testing centre. Patients may also be required to pay a share of the cost as not all tests are listed for reimbursement and reimbursement for some tests is limited to a ceiling.

Laboratory tests for screening purposes (*i.e.* performed in the absence of signs, symptoms, complaints or personal history of disease or injury) present additional issues. For example, in the United States, at present, such tests, which may include predictive/susceptibility tests, are not covered under Medicare (Centers for Medicare and Medicaid Services, 2004). As a result, Medicare's coverage criteria for pharmacogenetics-based diagnostics may be applied differently under different circumstances. For example, the plan could cover pharmacogenetic tests performed in the presence of signs, symptoms, or a personal history of disease or adverse reactions.

In essence, current reimbursement systems for genetic testing are not structured to reward innovation or the added value of tests designed to improve health outcomes or quality of life. Unless measures are taken to modify current reimbursement practices, it will be difficult for the diagnostics industry to respond favourably to the opportunities created by pharmacogenetics (see OECD, 2005b, for further discussion of issues related to reimbursement of genetic tests).

In fact, genetic testing may not simply help improve patient outcomes and the efficiency of health care. If appropriately implemented, effective testing may help reduce aggregate costs by targeting therapies to those who can benefit. Policy on reimbursement of genetic tests needs to be founded on a clear understanding of the utility of such tests as well as the costs and benefits of using such tests across the whole health-care system.

Premium pricing for improved effectiveness

In competitive markets, the price for a good or service is determined by the value that consumers place on the product and the costs of its production. For various reasons, many OECD countries have moved away from this type of pricing for health-care products and services. Nonetheless, pharmaceutical companies are likely to seek higher financial compensation or premium pricing when producing a drug guaranteed to work on a small population for which other drugs are ineffective or cause harmful side effects. Premium pricing may also be sought for the accompanying test. This strategy will be justified by the small market size, the reduced uncertainty about risks and benefits, and the high cost of development.

Decisions to allocate limited resources to a highly effective but costly treatment for a small patient population can be particularly complicated. They may be perceived as stretching the notion of solidarity in a public health insurance system; they may also create greater risk selection problems in private health insurance. In the case of orphan diseases, society has clearly signalled a willingness to pay additional subsidies to encourage innovation. However, health-care frameworks in many countries are not neutral about treatments administered at high cost to a small population, particularly if the treatment is sporadic (*i.e.* delivered once or twice in a lifetime) and its long-term benefits are seen as uncertain.

An example is Fabry's disease, an inherited metabolic disorder that affects one in 40 000 people (see Box 6.2). Fabryzyme® is a highly effective enzyme replacement treatment for this disease which costs approximately USD 200 000 a year per patient. Although the benefits of the treatment are evident, the costs can absorb a considerable amount of the health-care spending of a given system.

Box 6.2. Enzyme replacement therapy for Fabry's disease in Finland

Enzyme replacement therapy for Fabry's disease became available in 2001. In 2002, the pharmaceutical company that developed the therapy conducted a very small-scale clinical study in Finland, during which it provided free therapy for 13 patients.

In 2003/04, after the study had concluded, the hospital district responsible for these Fabry patients stopped treatment, arguing that the enzyme replacement therapy was too expensive (EUR 200 000 a year per patient) and the evidence of benefits was uncertain.

In response, the Finnish Fabry Society filed a complaint with the Parliamentary Ombudsman*, pointing out that the enzyme replacement therapy is provided in many other European countries. In its decision, the Parliamentary Ombudsman stated that the Finnish legislation does not allow some patients to be denied access to necessary treatment because of costs (see Decision 921/4/04 of the Parliamentary Ombudsman of Finland). Decisions should be solely based on the patient's need, the nature of the disease, and the effectiveness of the treatment. Treatment was thus continued.

In practice, costs are an important component when making decisions related to priority setting for allocating health-care resources in Finland. Enzyme replacement therapy for Fabry's disease creates what may be considered an "allocation dilemma".

***www.oikeusasiamies.fi/Resource.phx/eoa/english/index.htx*

Cases such as enzyme therapy for Fabry's disease create what is known as an "allocation dilemma". They require considering the – often competing – perspectives of patients, clinicians and society at large. Standard economic evaluation methodologies may struggle to deal with such cases as there is no agreed system to help decision makers make appropriate allocation choices (Danzon and Towse, 2002). Despite the availability of outcome measures, such as quality-adjusted life years (QALYs) gained, defining individual benefit is at an early stage of development. Mechanisms to attach value to that additional benefit – and to reflect the broader perspective of a society – are even less developed.

Such issues are brought into even sharper focus with pharmacogenetics-based medicines. In effect, pharmacogenetic developments present decision makers with two main challenges. First, they must find ways to encourage those who carry out economic evaluations to account for the differential effectiveness of technologies within different groups and genotypes. Second, they must consider whether and how the evaluation methodology itself might evolve to better assess the value of the additional health benefits. In some cases, benefits may not be clear. For example, they may depend on the methods used to define causal relations between a genotype and serious adverse reactions or to determine the sensitivity and validity of a biomarker test.

Cost effectiveness analyses and health technology assessments need to take account of these issues, and subsequent decision making needs to take account of these analyses and of the ethical, legal and social issues related to the impact of genetic testing and to the potential increase in therapeutic orphans (see OECD 2005a, Chapter 8, for further discussion).

Clearly, pricing and reimbursement schemes for pharmacogenetics present multiple challenges. The barriers are considerable and the extent to which payers are prepared to award higher prices for greater benefit is crucial. The ability to deliver pharmacogenetics in an environment in which cost is a consideration depends on broad agreement among specialists, and on the development of evidence-based guidelines that will contribute to clinical practice and reimbursement decisions. It will also critically depend on the development of reimbursement schemes that provide appropriate incentives to various components of the system, allowing for fair financial benefits.

Challenges for capturing the value of pharmacogenetics

An increasing number of OECD member countries are using economic evaluation techniques to guide decisions about health-care reimbursement (Dickson *et al.,* 2003). These techniques are most commonly used to assess the cost effectiveness of a new technology by comparing the costs and outcomes of alternative treatments and calculating the incremental cost-effectiveness ratio (ICER). This ratio is a simple expression of the cost per health outcome, such as life-year gained or a quality-adjusted life-year gained (QALY).

The next step for decision makers is to define what represents "good value for money" rather than "poor value for money". In most jurisdictions, cost-effective thresholds that determine good and poor value are set arbitrarily. They do not necessarily reflect the values that patients place on health outcomes; instead they represent the values that governments or their delegates place on additional health gains (Birch and Gafni, 2006; Gafni and Birch, 2006). For jurisdictions that are simultaneously concerned about equity and containment of health-care expenditure, some form of social decision making is inevitable. A number of countries are taking steps to make such decisions more reflective of the community's values, often through initiatives such as citizen councils and patient representation on decision-making bodies.

Empirical analyses carried out by the National Institute for Clinical Excellence (NICE) in the United Kingdom and the Pharmaceutical Benefits Advisory Committee (PBAC) in Australia provide some idea of how decision makers in these two institutions perceive good and poor value. A recent PCAB study found that between 1994 and 2003, the highest cost at which the Committee recommended a drug for listing was USD 52 400 per QALY. Above this value, nine (82%) of all applications were rejected or withdrawn by the manufacturer and two were given only conditional approval (subject to price reductions). This compares with rejection (or withdrawal) of 41 of 89 listing applications for which the estimated cost per QALY gained was below USD 52 000 (Harris *et al.,* 2006). Examination of the NICE recommendations suggests a threshold of acceptable cost-effectiveness of GBP 20 000-30 000 per QALY (Devlin and Parkin, 2004). This sends a signal to manufacturers about what decision makers may be willing to pay for improvements in mortality, morbidity or other cost savings.

However, recent literature identifies numerous limitations on the ICER threshold approach, some of which are pertinent to pharmacogenetics. The ICER is a static measure of a therapy's average incremental cost and average incremental health outcome (*e.g.* a QALY). In reality, clinical trial data often show widespread heterogeneity in costs and response among patients. A large standard deviation for costs and outcomes suggests that there will also likely be considerable variation in range of possible ICERs. This implies that the mean ICER may indeed be deemed good value, but there is a risk that the "true" ICER may be deemed poor value (see Box 6.3). Reliance on mean values alone may conceal the level of risk associated with reimbursement decisions and, as a result, undervalue drugs that offer lower risks.

Box 6.3. Reducing uncertainty in valuing pharmacogenetics

The incremental cost-effective ratio (ICER) is used to compare a new therapy to an old, as in Figure 6.2 in which the horizontal and vertical axes measure incremental health gain and costs, respectively. The mean incremental health outcome (*e.g.* a life-year gained) of a new therapy is shown by ΔE; the mean incremental cost by ΔC. These mean values provide an ICER through the origin of $ICER_m$. However, clinical data provides a measure of variation that produces a confidence interval (CI) for both costs and effects. These CIs define a lower and upper bound ICER (marked by $ICER_h$ and $ICER_u$ in Figure 6.2). Collectively, these elements outline an area, marked by "abcd" in Figure 6.2, on the cost-effectiveness plane within which one can be quite confident (95%) of locating the true ICER.

Figure 6.3 replicates Figure 6.2 but adds the ICER threshold value, labelled $ICER_t$. This is the value that decision makers have indicated as being good value for money. Any new therapy with an ICER below this threshold will be deemed good value; anything above it, bad value. Figure 6.3 is drawn to indicate that the mean ICER of the new technology is below the threshold value, *i.e.* it is good value. However, adopting this new technology carries a certain inherent risk. Because it is possible that the true ICER lies above the threshold ICER, decision makers face a small risk of adopting a bad value technology. The level of risk is marked by the triangle "aef".

The advent of pharmacogenetics creates the potential to reduce the uncertainty associated with patient response by making it possible to predict who will respond to a drug and who will not. This implies a narrower confidence interval (CI), especially in relation to the effect (or health outcome) of the new therapy. It also implies a narrower CI around the mean ICER and, therefore, less risk for decision makers who want to adopt cost-effective technology.

For reimbursement programmes that use cost-effectiveness analysis as an input in decision-making, the above discussion underscores the need for the routine reporting of uncertainty around the mean ICER.

Figure 6.2. Incremental cost effectiveness and confidence regions

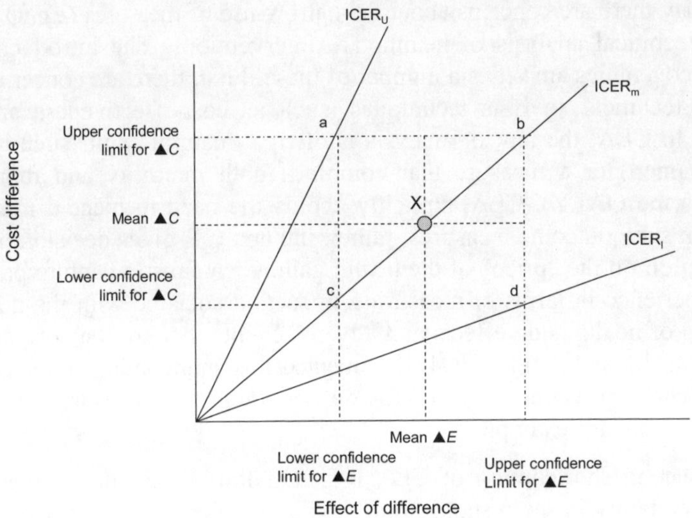

Figure 6.3. Incremental cost effectiveness and confidence regions and the threshold value

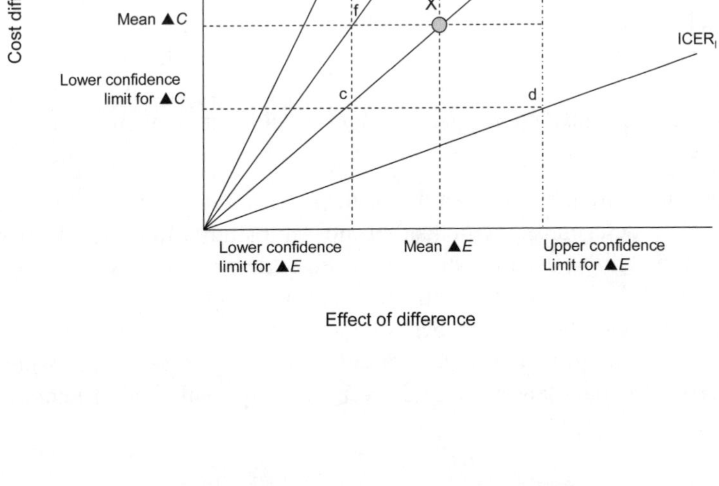

Using economic evaluation techniques in assessing pharmacogenetics

Today there are concerns about the narrow use of measures (*e.g.* QALYs) in the technical analysis of health-care interventions. The introduction of pharmacogenetics amplifies a number of these. First, there are concerns with current technical analysis techniques, such as cost-effectiveness analysis (CEA). In CEA, the unit of analysis is often a "natural" unit (such as life-years gained) or a measure that combines both mortality and morbidity (such as the QALY). CEA explicitly avoids the need to place a monetary value on such outcome measures; rather, the analysis gives decision makers information on the "price" of the health gain. CEA fails to capture patients' real experience in terms of measuring impacts associated with the duration of states of health, side effects and adverse events. An analysis of patients' views (Devlin and Parkin, 2004) documented this shortcoming and suggested that outcome measures used in CEA do not adequately measure the impact of side effects, for example.

A second shortcoming of CEA is that it fails to capture the value of greater certainty in the response rate: it captures the total number of health gains, irrespective of their distribution. Under certain scenarios, introducing a test to predict response will not increase the expected total health gain in the population. However, it will improve the ability to predict who is going to respond. For patients, this reduces uncertainty about whether or not they will respond to treatment. Under the assumption that patients are risk averse (*i.e.* they prefer less risk to more risk for any given expected return), it implies greater overall utility (*i.e.* society is better off).[7] This concept is illustrated numerically in Chapter 4, Box 4.2, in which the total health gain remains the same under two different scenarios but greater certainty of response to a drug may increase a patient's willingness to pay.

These concerns call for a search for additional techniques that more accurately capture patient values. Techniques such as willingness to pay and discrete choice analysis may provide decision makers with additional information regarding how patients value factors such as fewer adverse events and less risk in responsiveness. Such values could supplement evidence that derives from CEA.

Current approaches to health services research often preclude policy makers from determining which subgroups of a study might benefit from an intervention. This is usually because the study design emphasises measuring the effect of the intervention on an "average" population. Although less-advantaged populations often benefit the most from health-care spending, current standards in technological research fail to give decision makers the data needed to enact policies that deliver more equitable and efficient health outcomes.

The advent of pharmacogenetics calls for further methodological improvements in the conduct and reporting of economic evaluations to take account of the impacts of pharmacogenetics on the health system overall and in the long run, owing to more effective dosing and therapies and better patient choice. Thus economic evaluation – and the decisions made on that basis – need to be carried out within the framework of a more inclusive and far-sighted perspective.

Conclusions

Regulatory authorities have a clear interest in seeing the benefits of pharmacogenetics applied to both new and existing medicines. Moreover, they are in a position to influence the introduction of pharmacogenetics as a tool that will benefit human health and support innovation. The development of transnational regulatory approaches and policies, coupled with efforts to close existing gaps in relevant practices, can only increase this influence. One objective of such transnational dialogue might be to achieve consensus on the process by which regulatory authorities choose to approve and validate pharmacogenetic tests for a drug or to modify a label. Related but distinct issues around economic and other health technology assessment and post-marketing coverage and reimbursement decisions also call for transnational dialogue.

For new medicines, regulatory authorities are in a position to provide incentives that steer the pharmaceutical industry away from the blockbuster drug model and prompt it to focus instead on more targeted medicines. Without pressure from regulators, industry is unlikely to undertake major developments in this direction in the near future. To take an effective leading role, regulatory authorities need to consider whether it is sufficient to engage in early dialogue with companies and to encourage co-development of drugs and devices, or whether other steps will be required to enable a more radical approach to risk management and achieve an appropriate balance between patient benefit and potential harm.

For established medicines, regulatory authorities may represent the only major group (outside the research community) with any significant self-interest in applying pharmacogenetics. It is unlikely that pharmacogenetic tests are needed to assist in prescribing choices for common (and often over-the-counter) medicines such as aspirin and ibuprofen. However, such testing could contribute to therapeutic or dosing choices for a number of established medicines or could help to limit the risk of adverse drug reactions. In these cases, regulatory authorities have an opportunity to show significant leadership in terms of modifying labels on the basis of sound research and encouraging research to identify appropriate biomarkers for the many medicines for which such indicators have not been established. The question of who should bear the costs of generating such data (innovators, generic manufacturers, regulators or some kind of risk-sharing arrangement such as public-private partnerships) remains open, however.

Regulatory authorities across OECD countries have launched systematic reviews of marketed drugs to determine whether their labels should recommend genotyping to guide therapy. They are also encouraging systematic research on how specific genetic factors confer risk in relation to susceptibility to specific ADRs.

Pharmacovigilance is crucial for providing safety information that cannot realistically be collected prior to drug approval. It can help identify risk factors, reveal interaction with other medicinal products, and support the analysis and provision of information on the safe and effective use and regulation of medicinal products. Efforts to link pharmacovigilance systems to pharmacogenetics will depend on international agreement and co-operation, the establishment of appropriate incentives, and broad acceptance on the part of physicians and patients.

Reimbursement systems are a strong factor in the rate of diffusion of pharmacogenetics. In flexible systems, reimbursement decisions can be used to send signals to industry about the price levels considered acceptable for specific clinical gain. The use of economic evaluation and evidence-based policies for drug coverage could go some way towards achieving pricing that better reflects benefits.

However, mechanisms to attach value to benefits – and to reflect the broader perspective of a society – are not well developed. In most jurisdictions, cost-effective thresholds that determine good and poor value are set arbitrarily. They do not necessarily reflect the values that patients place on health outcomes; rather, they represent the values that governments or their delegates place on additional health gains. Hence, decision makers face two main challenges with regard to pharmacogenetics. First, they must find ways to encourage those who carry out economic evaluations to account for the differential effectiveness of technologies within different groups and genotypes. Second, they must consider whether and how the evaluation methodology itself might evolve to better assess the value of the additional health benefits. International dialogue and co-operation across sectors (public and private) will be necessary to develop such methodologies.

In most OECD countries, current reimbursement systems for genetic testing are not structured to reward innovation, nor do they recognise the added value of tests designed to improve health outcomes or quality of life. Unless measures are taken to modify current reimbursement practices, it will be difficult for the diagnostics industry to respond favourably to the opportunities created by pharmacogenetics.

The barriers to the adoption of pharmacogenetics are considerable, and the extent to which payers are prepared to award higher prices for greater benefit is crucial. In an environment in which cost is a key consideration, the ability to deliver pharmacogenetics depends on three factors:

- broad agreement among specialists;
- the development of evidence-based guidelines that contribute to clinical practice and reimbursement decisions;
- the development of reimbursement schemes that provide appropriate incentives to various components of the system and allow for fair financial benefits.

Notes

1. UK Clinical Research Collaboration, *www.newscientistjobs.com.*

2. *http://nihroadmap.nih.gov/clinicalresearch/.*

3. For recent reviews see Eurobiobank (2005), Knoppers and Scriver (eds.) (2003), Martin and Kaye (1999) and CIOMS (2006).

4 For example, as supported by the UK Department of Health, *www.genres.org.uk.*

5. "Table of Valid Biomarkers in the Context of Approved Drug Labels", *www.fda.gov/cder/genomics/genomic_biomarkers_table.htm.*

6. Announcement of the Collaborative Cardiovascular Drug Safety and Biomarker Research Program, *www.fda.gov/oc/initiatives/criticalpath/biomarker.html.*

7. For example, a person is given a choice between a 50% chance of winning USD 20 and a 75% chance of winning USD 15. The expected return is the same in each case (*i.e.* 0.4*USD 20 = 0.8*USD 10 = USD 8) but a risk-averse person will prefer the second because less risk is involved.

References

Bakke, O.M., M. Manocchia, F. de Abajo, K.I. Kaitin and L. Lasagna (1995), "Drug Safety Discontinuations in the UK, US and Spain from 1974 to 1993: A Regulatory Perspective", *Clinical Pharmacology and Therapeutics*, 58, pp. 108-117.

Birch, S. and A. Gafni (2006), "The Biggest Bang for the Buck or Bigger Bucks for the Bang: The Fallacy of the Cost-effectiveness Threshold", *Journal of Health Services & Research Policy,* 11(1), pp. 46-51.

Centers for Medicare and Medicaid Services (2004), *Medicare National Coverage Determinants (NCD) Coding Policy Manual and Change Report*, Centers for Medicare and Medicaid Services, Baltimore, MD.

CIOMS (Council for International Organizations of Medical Sciences) (2006), "Management of Safety Information from Clinical Trials", *www.cioms.ch/frame_management_of_safety_information.htm.*

Danzon, P. and A. Towse (2002), "The Economics of Gene Therapy and of Pharmacogenetics", *Value in Health*, 5(1), pp 5-13.

Devlin, N. and D. Parkin (2004), "Does NICE Have a Cost-effectiveness Threshold and What Other Factors Influence its Decisions? A Binary Choice Analysis", *Health Economics* 13(5), pp. 437-52.

Dickson, M., J. Hurst and S. Jacobzone (2003), "Survey of Pharmacoeconomic Assessment Activity in Eleven Countries," OECD, Paris.

Eurobiobank (2005), "Outstanding Legal and Ethical Issues on Biobanks: An Overview of the Regulations of Member States of the Eurobiobank Project", *www.eurobiobank.org.*

Gafni, A. and S. Birch (2006), "Incremental Cost-effectiveness Ratios (ICERs): The Silence of the Lambda", *Social Science & Medicine*, 62(9), pp. 2091-2100.

Harris, A.S. and G. Chin (2006), "Uncertainty, Economics and the Decision to Fund Drugs in Australia 1994-2004", *Health Technology Assessment International*, Adelaide.

Knoppers, B.M. and C. Scriver (eds.) (2003), *Genomics, Health and Society: Emerging Issues for Public Policy*, Public Research Initiative, Canada, *http://policyresearch.gc.ca/doclib/genomicbook_e.pdf.*

Martin, P. and J. Kaye (1999), "The Use of Biological Sample Collections and Personal Medical Information in Human Genetics Research: Issues for Social Science Research and Public Policy", Wellcome Trust Report, Wellcome Trust, London.

OECD (2005a), *Health Technologies and Decision Making*, OECD, Paris.

OECD (2005b), *Genetic Testing: A Survey of Quality Assurance and Proficiency Standards*, OECD, Paris.

OECD (2006), *Creation and Governance of Human Genetic Research Databases*, OECD, Paris.

OECD (2007), *Guidance on Best Practices for Quality Control of Human-derived Biological Resources*, OECD, Paris.

Pirmohamed M. *et al.* (2004), "Variability in response to Warfarin®: a prospective analysis of pharmacogenetic and environmental factors", *www.genres.org.uk/prp/projectliverpool2.htm.*

Rägo, L. (2003), *WHO Drug Information* 17, pp. 84-91.

Shah, R.R. (2004), "Drug Development and Use in the Elderly: Search for the Right Dose and Dosing Regimen", *British Journal of Clinical Pharmacology*, 58, p. 452.

Shah, R.R. (2006), "Can Pharmacogenetics Help Rescue Drugs Withdrawn from the Market?", *Pharmacogenomics* 7, pp. 889-908.

Uyama, Y. (2005), "Perspective and Strategy on Pharmacogenetics", paper presented at the OECD workshop "An International Perspective on Pharmacogenetics: The Intersections between Innovation, Regulation and Health Delivery", Rome, October, *www.oecd.org/sti/biotechnology/pharmacogenetics*

WHO (2002), Genetic Databases – Assessing the Benefits and the Impact on Human and Patient Rights, *www.law.ed.ac.uk/ahrb/publications/online/whofinalreport.pdf.*

Chapter 7

Conclusions

While research in pharmacogenetics is proceeding rapidly, at the start of 2009 there were only just over a dozen pharmacogenetics-based products on the market. A number of scientific, economic and regulatory challenges need to be overcome if pharmacogenetics is to be taken up more widely within health-care systems. It is not yet clear to the private sector what business models will deliver acceptable returns on investment. In the public sector, governments have a number of levers which they could use to create an "enabling" environment for pharmacogenetics, while continuing to provide the necessary checks and balances.

This report identified six areas on which public policy and co-ordinated international action might focus. These include:

- developing the infrastructure necessary to identify and validate the genetic biomarkers that underpin pharmacogenetics;
- building public-private collaborations in order to apply pharmacogenetics to established medicines, thus delivering broad public benefits through improved safety and efficacy for patient populations;
- aligning incentives so that the private sector is encouraged to invest in pharmacogenetics and transform the drug development process;
- supporting actions by regulatory authorities that incrementally set the future policy environment and also influence investment decisions regarding pharmacogenetics;
- assessing the health and economic impacts of applying pharmacogenetics to health technologies. This will require an evolution in the models and methodologies used as well as agreement on how and by whom the necessary underpinning data will be developed;
- supporting effective point-of-care delivery of pharmacogenetics-based technologies. Active evidence-based policies will be needed to bring such technologies into health-care systems.

Research infrastructures

Scientists' understanding of the response of individuals and of sub-populations to diseases and drugs at a genome-wide scale is progressing extremely rapidly. The biomedical research community is developing a large body of knowledge about the interactions between genes and diseases and drugs; potential biomarkers are being identified and a subset of these is developed into validated clinical tests which are utilised by health professionals.

However, the broader uptake of genetic biomarkers faces a number of hurdles. During the research and development phase, a major problem is the expense and time necessary to access, link and integrate the necessary genetic data and related health data. Another is reaching agreement on the type of data to be collected in order to establish, ultimately, the clinical utility of a biomarker. A third problem – further downstream – is the funding and performance of any large-scale patient cohort studies necessary for clinical evaluation and validation.

To succeed in delivering this new model for medicine development, more "open" approaches to sourcing data and materials will be essential. As discussed in Chapter 2, countries have some options for encouraging sharing and collaboration, such as:

- encouraging the development of mechanisms that make basic pharmacogenetics data more easily accessible and shared;
- supporting the creation and utilisation of large-scale human biobanks and genetic research databases (HBGRDs);
- encouraging the formation of multidisciplinary international networks which can increase the efficiency of pharmacogenetic research;
- fostering the development of systems to manage knowledge and intellectual property and support open innovation platforms for pharmacogenetics;
- considering the formation of public-private partnerships to carry out association studies.

Such infrastructure should not be developed in a piecemeal fashion; interventions need to be co-ordinated and coherent. Pharmacogenetics – and the informatics needed to harness its power – can make a continuous multifaceted cycle of innovation a reality. The linear policy model needs to be discarded, and policy development needs to reflect this change in the system of innovation.

Improving drug development

There is a growing consensus that new tools and approaches are needed to improve decision making during the drug discovery and development process and some high-profile initiatives have been launched to help achieve this. Significant savings might be realised, for example, by identifying, as early as possible in the development process, lead molecules that are likely to fail to become effective medicines and discard them. Pharmacogenetics also shows potential for generating other efficiency gains. In particular, biomarkers are increasingly used as surrogates for outcome measures or clinical endpoints; this decreases the number of human subjects needed in a clinical trial and the time necessary to demonstrate the significance of the effect of a drug. Molecular profiling methods (DNA micro-arrays and proteomics) can be used to detect molecular changes due to disease, drug treatment or toxicity. Pharmacogenetics, thus used, can improve the efficiency of the drug development process by identifying more quickly which candidates are likely to succeed or fail.

Moreover, pharmacogenetics allows stratification of patients in clinical trials into "responders" and "non-responders". Using genetic screening to identify potential responder populations prior to clinical trial enrolment would make it possible to demonstrate drug efficacy in a smaller set of subjects. This approach could, in principle, both reduce the overall costs of clinical trials for a prospective medicine and increase the probability of the success of individual trials. There are uncertainties though about the long-term impact of these developments. The cost and duration of trials will depend, for example, on how many patients are needed to obtain statistically meaningful data on the predictive response profiles. And while enrichment (*i.e.* focusing only on the target patient group) can be advantageous, there may also be a greater risk that rare or delayed side effects may go undetected if trial population sizes are small.

With pharmacogenetics, drug development is being transformed into a data-driven process in which the four clinical phases largely overlap. Pharmacogenetics in this way too is making a reality of the conceptual "virtuous cycle" of innovation and promises to bring to an end the inefficient and unrealistic linear model of innovation that pervaded the late 20th century.

Business models

Widespread adoption of an approach to drug discovery and development that draws on the power of pharmacogenetics and offers more targeted interventions is largely dependent on the private sector's perception of the financial opportunities it offers. Understanding the incentive environment for health innovators is therefore vital to harnessing the benefits of advances in pharmacogenetics research for public health.

Widespread adoption of pharmacogenetics is likely to mean that new business models will be developed in the pharmaceutical and diagnostics sectors. By enabling market stratification (either by drug or by patient group), pharmacogenetics may, in fact, restrict the market size for a drug. Perspectives diverge as regards the extent to which a shrinkage in market volume would be offset by an increase in the prices paid for better targeted therapies; this report takes the view that a reduction in market size is likely to put pressure on the aggregate value of markets for individual therapies as well as reduce volumes. Instead of relying on a single blockbuster drug, pharmaceutical companies may need to develop a more diversified portfolio of drugs that are individually less profitable than a "blockbuster" but collectively may yield greater health benefits. Some are sceptical that the pharmaceutical sector will willingly embrace this change, especially while the financial yields remain uncertain. However, factors that could push towards the early adoption of targeted therapies include: *i)* reduced attrition and more success in developing pharmacogenetic drug candidates; *ii)* the opportunity to pursue drugs previously discarded during the development phase or withdrawn post-marketing because of toxicity in specific patient sub-groups or uncertainties regarding efficacy; and *iii)* the balances struck among changes in the regulatory, evaluation and reimbursement systems – and the timing of these – in response to the availability of pharmacogenetics-based evidence.

For the device industry to invest in pharmacogenetics, diagnostic firms will need to be able to capture more value than at present, or they will need to form a closer relationship with pharmaceuticals developers. As diagnostics are typically developed after a target has been identified, diagnostics companies have little involvement in target identification. The intellectual property protection afforded diagnostic devices is weaker, and harder to enforce, than for new chemical entities. Moreover, most pricing and reimbursement schemes for diagnostics are cost-based and may not adequately reflect the value of a testing device's contribution to the patient or to the health-care system. As a result, profits from diagnostics are based on volume of sales, and competition arising from the rapid development of better or cheaper devices is intense.

Many believe that pharmacogenetics will require a business model in which drugs and devices are developed and perhaps marketed in parallel. The co-ordination of these activities could take a number of forms, from alliances, to mergers and acquisitions, to expansion of pharmaceutical companies' in-house expertise in diagnostics. In such a co-development and co-marketing model, the pharmaceutical and diagnostic products would capture the value from their innovations together. Others feel that stand-alone diagnostic companies could remain players in pharmacogenetics, but that current pricing and reimbursement mechanisms will need to change if there are to be stronger incentives to develop new, genetics-based assays.

The absence of validated genetics-based assays could stymie the effective use of pharmacogenetics-based evidence and the uptake of targeted therapies. Action on the part of governmental and regulatory bodies as well as private health insurers may therefore be needed to catalyse change in areas such as pricing and reimbursement; recognition of the value of diagnostic tests to the health-care system as a whole; and, possibly, mechanisms for protecting intellectual property and better capturing value from biomarkers.

Regulatory challenges

Regulatory authorities, and society in general, have a clear interest in seeing pharmacogenetics data used to assess the safety and efficacy of both new and existing medicines. Regulatory actions will have a strong influence on incentives to introduce pharmacogenetics-based products. What is perhaps particularly important is the process by which regulatory authorities choose to approve and validate pharmacogenetic tests for a drug or to modify its label.

However, the validation of genetic biomarkers for use in the regulatory approval process is still subject to considerable uncertainty. A recent trend among regulatory authorities is to ask that pharmaceutical companies voluntarily indicate which patient populations are likely to benefit from a new treatment. Regulatory requirements to demonstrate the greater efficacy and cost-effectiveness of drugs are also prompting industry to provide more pharmacogenetic data. However, while regulatory agencies accept pharmacogenetic data (and provide guidance on their submission), they do not as yet require them. If regulatory agencies are indeed moving toward requiring such data, discussions by innovators and regulators will be needed to clarify the sort of data submission needed and how it will be used in the approval process.

In the meantime, regulation of the combined use (co-development) of a therapeutic with a diagnostic is evolving across OECD member countries. There is concern that pharmacogenetics will make an already complex approval process for pharmaceuticals even more complicated. How regulation will deal with the co-development of drugs and diagnostics needs to be clarified in terms of data requirements for approval and of how reimbursement systems will react to co-marketing of co-developed products.

While pharmaceutical companies see the value of investing in the development of biomarkers for new drugs, they have no incentive to apply pharmacogenetics to established medicines (particularly those that are off-patent). Pharmacogenetic tests are unlike to be needed to assist in prescribing choices for common or over-the-counter medicines such as aspirin and ibuprofen but they can aid therapeutic or dosing choices for other established medicines. In such cases, regulatory authorities have an opportunity to show leadership by modifying labels on the basis of pharmacogenetic research and encouraging research to identify appropriate biomarkers for the many medicines for which such indicators have not been established.

Prospective studies would allow regulatory authorities to determine whether it is necessary to modify the labelling of specific established medicines in order to improve clinical practice and patient health. However, as the cost of such trials may be very high, the application of pharmacogenetics to established medicines becomes an issue of public policy. It is inequitable, and inefficient in terms of the stability of market incentives, to place the full burden of retrospective genomic analysis on the holders of proprietary rights. Ways need to be found to share the burden and risks associated with such analysis.

OECD countries may want to consider the options of conditional market approval, of post-market pharmacovigilance, and of risk-sharing mechanisms for certain products based on pharmacogenetics evidence. Pharmacovigilance provides safety information that cannot realistically be collected prior to drug approval. It can help identify risk factors, reveal interactions with other medicinal products, and support the analysis and provision of information on the safe and effective use and regulation of medicinal products. There is an opportunity to use pharmacovigilance systems to contribute to *ex post* evaluation of assumptions based on pharmacogenetics evidence, but this would require appropriate incentives and broad acceptance on the part of physicians and patients.

Reimbursement policies and assessment methodologies

The economic incentives to invest in the development of biomarkers are influenced by signals that flow from the broader health-care system. In most OECD countries, the current reimbursement systems for genetic testing are not specifically structured to reward innovation, nor do they recognise the added value of tests designed to improve health outcomes or quality of life. Changes to reimbursement policies could help make the development of new, genetics-based assays more attractive. For example, clear signals from governmental bodies about how this technology will be priced and reimbursed, recognition of the added value of diagnostic tests for the health system as a whole, and mechanisms for capturing and protecting the intellectual property inherent in diagnostic tests might improve the climate for investment.

Studies on the health economics of pharmacogenetics and on the cost-benefit ratio of pharmacogenetic testing and products are necessary to provide the evidence base for their uptake. There is at present a lack of data to demonstrate the clinical utility and cost-effectiveness of many pharmacogenetic therapeutics and diagnostics, and there are debates about the sort of information needed and no agreement about the responsibility for developing such data. It is likely to be necessary to develop new models or methodologies for assessing diagnostics and medicines, as well as sharing data, which could eventually influence the pricing and reimbursement of such products.

Health-care systems

Pharmacogenetics already contributes to better clinical care by improving our understanding of disease heterogeneity, reducing the uncertainties of responses associated with specific treatments, changing the risk-benefit ratio for treatments, and enhancing the ability to prescribe accurate dosage for medicines.

However, pharmacogenetic assays rarely provide simple, clear-cut results: the information generated is probabilistic rather than absolute. Before pharmacogenetic testing becomes a routine part of health care, more evidence needs to support its clinical utility, information on pharmacogenetics and other relevant information need to be available at the point of care, and health-care providers need to be educated and trained to access and interpret these new sources of health data.

Successful, large-scale uptake of pharmacogenetics by clinicians will depend on:

- evidence about the clinical utility of pharmacogenetic testing;
- the ability to integrate information across platforms and communicate among multiple end-users within a health-care setting;
- availability of pharmacogenetics and other relevant information at the point of care.

Current applications suggest that pharmacogenetics is unlikely to "revolutionise" health-care systems in the near future. It is, however, generating a body of data that will help transform medical care from a reactive system (*i.e.* one that focuses on episodic treatment) to a proactive system that can provide more personalised medicine.

Social choices

Evidence shows that patterns of genetic variation can be identified with specific subpopulations and ethnic groups and may affect that group's response to a medicine. Supportive approaches will be needed to extend pharmacogenetics research to these specific ethnic groups, particularly for well-established essential medicines.

Society will need to decide how to mitigate risks of increasing genetic exceptionalism or inequity in access once pharmacogenetics enables stratification of patients into different groups (*e.g.* responders and non-responders). Debates about the equity of health-care delivery are to be expected, particularly in terms of possible exclusion from clinical care that involves potentially life-saving drugs.

For the most part, patients do not distinguish between pharmacogenetic drugs and testing and more traditional approaches to medicine. However, the public has expressed concern about a number of ethical, social and legal issues pertaining to genetic testing. These issues relate to the storage and handling of genetic samples and data; the nature of the consent required; and the end use of the data extracted from genotyping, particularly its potential release to third parties. Governments need to continue to address these concerns as genetic testing becomes more common and the data derived are used in an expanding variety of settings (*e.g.* research and clinical). The way in which these issues are addressed will affect public attitudes and trust in genetics-based medicines.

OECD PUBLISHING, 2, rue André-Pascal, 75775 PARIS CEDEX 16
PRINTED IN FRANCE
(93 2009 08 1 P) ISBN 978-92-64-07679-2 – No. 57041 2009